MERCURY IN THE ECOSYSTEM

Its Dispersion and Pollution T

AF734884

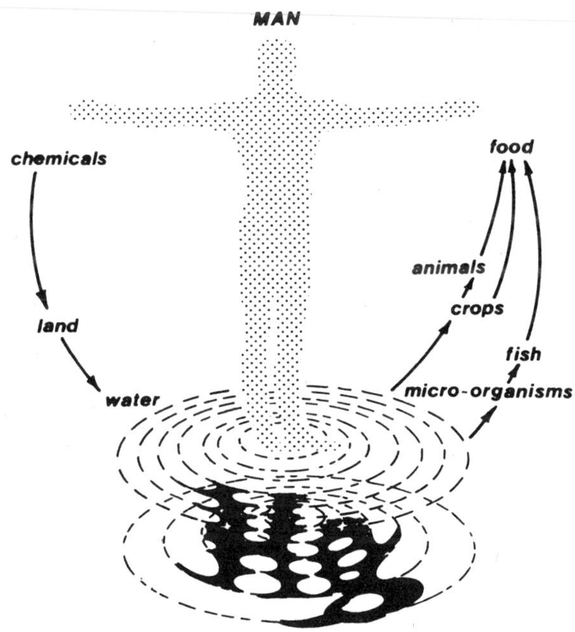

Sachinath Mitra

M. Sc., Ph. D., D. I. C.

Professor of Geology, Jadavpur University
Member, Telephone Advisory Committee, Calcutta
Ex-Jt. Director, Planning Commission, New Delhi
Visiting Professor, Bhabha Atomic Research Centre, Bombay
Visiting Scientist, NASA Johnson Space Centre, Houston.

TRANS TECH PUBLICATIONS 1986
Switzerland - Germany - UK - USA

the MANCHESTER
METROPOLITAN
28 JUL 1999
UNIVERSITY LIBRARY

Copyright© 1986 Trans Tech Publications Ltd., Switzerland
ISBN 0-87849-529-0

Distributed in the Americas by

Technomic Publishing Co., Inc.
851 New Holland Ave.
Box 3535
Lancaster PA 17604
USA

and worldwide by

Trans Tech Publications Ltd.
Trans Tech House
4711 Aedermannsdorf
Switzerland

Printed in the Netherlands

C 553·454 MIT

Dedicated to my late Father

Debendra Nath Mitra

'Life is an honest try'

D.N. MITRA

CONTENTS

FOREWORD

Mercury occupies a unique place in the field of chemistry. It was the idol of the alchemists of ancient times, and throughout the ages they tried to convert it into gold. It was the metal which was first used for demonstrating the presence of oxygen in air. This metal, particularly its salts, has been extensively used by various life scientists as a potent inhibitor of bacterial growth. Its well-known marked affinity for the sulphydryl group, particularly in the proteins, has been used as a tool in understanding protein structures and functions.

During the last two decades, mercury has played a pioneering role in increasing the awareness of scientists, as well as the lay public, with regard to the consequences of industrial pollution leading to a disturbance of the ecosystem. The tragic case of mercury poisoning, which caused the horrible affliction known as "Minamata Disease", was an eye-opener for scientists, industrialists, and politicians as to the consequences of the indiscriminate use of mercury and also the discharge of toxic metals into the environment. Even today, the undersigned vividly remembers the most gruesome image of a simple fisherman from a peaceful and quiet Japanese village - a victim of this metal.

The fact that mainly fishermen, rather than the whole population of the village, were affected provided a clue to the scientists, and investigations led to the detection of the culprit - the fish. The latter metabolized the mercuric pollutants into organic methylmercury, and this was the primary causative agent of the disease. This, and the incidence of mercury poisoning in Iraq due to the indiscriminate use of mercurial fungicides led to an awareness of the toxicity of this metal. Therefore, the discharge of mercury into the environment was strictly controlled, since it could cause a disturbance of the ecosystem, and very intensive studies were carried out in various parts of the world where mercury was used in several industries. It is indeed very surprising that, although we were aware of the toxic effects of mercury, we woke up to the need to control mercury pollution only after such incidents of human tragedy.

This monograph on Mercury in the Ecosystem is very timely. It is written by Professor S.Mitra, who is well-known in India for his multi-disciplinary research activities and multi-faceted approach to any scientific problem which he decides to tackle.

This comprehensive treatise points out the wide distribution of mercury in the geological environment. The use of mercury in various industries, and also the accumulation of mercury in the animal and plant kingdoms due to the increasing discharge of mercury into the ecosystem as various types of industrial effluent, have been documented in the monograph. Various analytical methods for estimating the content of mercury in organic and inorganic matter have been described herein, and these will be of considerable help in selecting the proper techniques to be used for the estimation of the concentration of this toxic metal. The effects of environmental pollution by this metal upon the human population have been discussed in great detail, and the symptoms of, and remedies for, mercury poisoning have been described in a lucid manner. An attempt has also been made to present recent concepts regarding the mechanism of methylmercury formation in contaminated matter. The possible mechanism of its effect at the chromosome level has also been discussed.

I feel that this detailed review of Mercury in the Ecosystem is of paramount importance, particularly where rapid industrial development is in progress, because policy-makers all over the world still do not appreciate the essential need to take preventative measures. I hope that this monograph will influence mercury pollution control measures and help to avoid the recurrence of human tragedies. That major efforts should be directed towards mercury pollution control is obvious.

Professor B.K.Bachhawat, Ph.D., F.N.A., F.A.Sc.
Director,
Indian Institute of Chemical Biology,
4, Raja Subodh Mullick Road,
Jadavpur,
Calcutta - 700 032

January, 1984

PREFACE

The planet Earth's environment is dynamic and ever-changing. The influence of Man has been so great that virtually no pristine conditions exist to-day, not even in the upper stratosphere or in the depths of the ocean. By his very nature, Man has become a notorious polluting creature and the most injurious to the environment.

Chemicals have been used since the dawn of civilisation, although their use has increased phenomenally with spreading industrialisation. The huge consumption of chemicals has resulted in widespread contamination of the environment. The most widely-used compounds added to the environment are DDT-type insecticides, polychlorinated biphenyls (PCB's) and mercury compounds. They are now distributed around the world; even in the southern polar region. Indians are reported to have the highest concentration of DDT in their blood. Insecticides and industrial pollutants have been found in the tissues of penguins and fish in the Antarctic, and of whales in the Arctic Ocean. The fat of penguins contains mercury which has been fixed via the inhalation of atmosphere-transported mercury compounds. Remote glacial ices in Antarctica and in the Arctic also contain them. This reflects the global nature of mercury contamination via atmospheric circulation and other natural processes.

"Unpolluted" air contains less than 1, to about $10 ng/m^3$, of mercury. "Natural pollution", caused by the volatilisation of mercury from ore deposits of mercury or base metals contributes up to $62 ng/m^3$. Very large amounts of mercury reach the atmosphere from volcanic gases, fumeroles etc.

Mercury becomes globally dispersed throughout the atmosphere and its residence time in the latter varies from 5.5 to 90 days. Over a long geological period, the total mercury concentration in the oceanic waters remained constant, but increasing human activity has led to an increased dispersion of mercury. This gradually becomes more concentrated, by a factor of between two and four, in lakes and rivers. Sediment cores from lakes and estuaries indicate that there are two to five times higher levels of mercury in recent sediments compared to those in sediments laid down in pre-human culture times (UK Department of the Environment, 1976).

However, 25 to 30% of the atmospheric burden of mercury comes from anthropogenic sources, e.g., mining, earth-moving, burning of fossil fuels (coal and petroleum), ore smelting (for lead, zinc, copper, etc.), chemical industries (like chlor-alkali, paper, pulp, paints, etc.), insecticides (fungicides, pesticides, etc.), electrical and measuring equipment (thermometers, barometers, DC meters, pressure gauges, etc.), amalgamation and dental use, vinyl-chloride plastics, etc., etc. About 20000 tons of mercury per year are released by industry and this eventually becomes distributed in water (ground-, river-, and sea-); chapter III. Over the past one hundred years, anthropogenic activities have led to the discharge of about 100 million kilograms of mercury into natural waters, and about five times this amount each into the land and atmosphere. The amount in soil has also increased by about 0.02%. The annual mercury consumption of each country indicates that mercury losses into waterways, chemical dumps, air, etc., are related to the gross national product (GNP) of that country by:

$$\text{Hg consumption (thousands of 76lb flasks)} = 14.9\text{GNP}(10^{12}\text{US\$}) + 0.54$$

In this age of ecological panic, it is impossible to forego spontaneously the use of hazardous chemicals, and the threat has to be reduced through a better understanding of the factors involved in their circulation through the ecosystem. The price of a healthy environment is eternal vigilance.

In this monograph, several chapters have been devoted to a more detailed description of the geological formations, distribution, and dispersion of mercury than was originally thought to be necessary in order to equip the serious reader in environmental science with the required basic technical information. This was done while keeping in mind the lamentation of Frank Fraser Darling, the learned ecologist, in 1970 that, despite the prominence of ecology as a public issue and newsworthy subject, its prominence was built upon a remarkably weak 'educational' foundation.

The Earth's own reservoir of mercury in various rocks, mineral-types, and geological states can be meaningfully comprehended only through an 'educational' approach to the geological, mineralogical, geochemical and tectonic environment (chapter II).

The mercury contents of all analysed meteorites (falls) show an increasing trend as a function of their terrestrial age. Such a trend also indicates a high level of pollution of the atmosphere.

The three major components of the ecosystem: air (atmosphere), water (hydrosphere), and rocks and soil (lithosphere), control our planetary life. They are inextricably interwoven so as to support the fragile system of the living world. Toxic materials cannot be put into either air, water, or soil without their reappearing in the other two. The residence times of mercury in them are: atmosphere - 11 days, soil - 1000 years, oceans - 2100 years, and sediments - 2.5×10^8 years.

The major sources of mercury in the environment (chapter III) and its cycle or movement (chapter IV) are discussed, followed by a review of the analytical techniques employed (chapter V) and aspects of mercury pollution which affect mammals (chapter VI), plants and fish (chapter VII), and the nature of mercury poisoning and its therapy (chapter VIII). Biological amplification of pesticides and metal concentrations along the food chain are well-known phenomena. Large predatory fish concentrate 1000times more mercury in their flesh by eating small fish containing lower concentrations.

The concentration of DDT in bay shrimps in a Long Island estuary was only about 0.16ppm, but it rose to a massive 75.7ppm in the predatory gulls of the area. Elsewhere, there are predatory birds in which the DDT concentration has been found to be one million times as high as that in the waters where they obtain their food. Non-degradable insecticides (both mercurial and DDT-type) sprayed in the fields, or seeds coated with mercury in order to prevent fungal growth, may enter the food chain in a variety of ways. These not only increase the amounts of mercury in human beings but, in the process, the sudden eradication of one predator from the food chain may bring about a 'population explosion' of another pest or prey lying at another point along the chain.

Of all the compounds of mercury, the methyl and other alkyl forms are the most toxic to man. Almost all (98%) of the mercury found in animal tissues is in the form of methylmercury.

Methylation of inorganic mercury takes place in soil and sediments. In freshwater sediments, the methylation of inorganic mercury ranges from 5 to 15% per gram per year, and demethylation rates are of the same order (Spangler et al, 1973). On the other hand, a delicate balance of methylation and demethylation is maintained in nature by bacterial action. The biological cycle is delicately balanced, and small perturbations in the input and environment can result in a change in the bioenergetics, resulting in a change in the rates of methylation in sensitive ecosystems. Bacterial demethylation occurs over a broader range of redox conditions.

The most intense mercury methylation occurs in the transitional oxidizing anaerobic zones of natural sediments. Normally, methylation takes place in bottom sediments and is restricted to the upper 5 to 10cm. In water, fish and other biota are taken to be the ultimate sink for methylmercury.

Although the present emphasis is on protecting the human population from ingesting alarming levels of methylmercury in its diet, the neurological, teratogenic, and mutagenic effects of mercury have caused great concern for the impact of mercury compounds on future generations of men and animals.

In the USA and the USSR, a permissible limit for mercury (methylmercury) has been established for foodstuffs, and 0.05mg of Hg vapour per m^3 of air is the upper limit for industrial atmospheres. In drinking water, this limit has been set at the higher limit of 50ppb. India has yet to adopt similar safety standards for mercury usage and discharge (chapter X).

However, all over the world the use of organomercurials as insecticides has decreased greatly and they are still applied in small amounts (circa 2g/ha mercury) to cereal seeds. Alkylmercurials have been replaced almost completely by the less toxic phenyl or alkoxy mercurials. In some countries (USA, Japan and Sweden but not the UK or France), the use of even the latter groups has been severely restricted or banned over the last decade. However, in Sweden, mercurial treatments have been reintroduced in situations where there is a high risk of disease. In 1974, world sales of copper fungicides amounted to Ł83M, and mercurials to Ł23M (Anon, Farm Chemicals, 38, September, 1977) and sulphur contributed about Ł30M. However, Japan was able to reduce the consumption of mercury by 21% in 1975.

Substitutes for mercury have been found in many formerly important applications, and considerably less of it is now used for amalgamation (largest use in 1870's), antisyphilitics (largest use in 1917), munitions, detonators (fulminates), felt manufacture, pattern making, paper manufacture, vermillion, pyrotechnics, and cosmetics. In some major areas, substitutes are being tried but this often results in a downgrading of the quality of the product, or economy of manufacture, or both. For example, in the manufacture of chlorine using the caustic soda process, a diaphragm cell can be used instead of the mercury electrolytic cell, but it yields a product of lower purity at a higher cost. This ultimately affects the price of nylon, rayon, petroleum products, aluminium, detergents, and many other consumer items. However, for many applications such as batteries, mercury lamps, wall switches, thermostats, barometers, manometers, thermometers, pressure gauges, and other laboratory uses, substitution is essentially impractical. For agricultural applications and special paints, substitution may be possible in the future as non-mercurial germicides and fungicides having selective toxicity are discovered.

However, no major political party in the world ever promises the electorate that it will carry out its environmental responsibilities by shifting the emphasis from economic growth (which by its very nature contributes to pollution) to a genuine effort to protect the environment. Within the system, the citizen pays either as consumer, tax-payer, or victim.

The biotic and non-biotic environments are intertwined like the warp and weft of a carpet. The study of the ecosystem vis-a-vis any toxic element, therefore naturally violates the traditional academic boundaries of differing scientific fields (viz., zoology, geology, biochemistry, medical science, etc.) and becomes woven into the fabric of the subject known as ecology. Here, the zoological, biochemical and medical aspects of the study were taken care of by Dr Ashima Mitra, M.B.B.S., who utilised the library and other resources of Nil Ratan Sarkar Medical College, the Indian Institute of Chemical Biology, and the Department of Pharmacy, Jadavpur University, Calcutta.

In this work, considerable aid in literature searching and referencing, and help in the development of a mercury detector were provided by Dr P.K.Banerjee, Dr P.Ghosh Dastidar, Dr J.Datta, and Mr Malay Mukherjee, all of whom worked on the author's research projects between 1973 and 1979. The help of my son, Punyashlok Mitra, in checking the typescript is gratefully acknowledged.

The writing of this monograph was initiated under a Book Writing Scheme sponsored by the University Grants Commission, New Delhi. The encouragement of Dr D.Shankar Narayan, Secretary of the U.G.C., throughout the writing of this book is thankfully acknowledged.

Finally, the provision of library facilities by the following institutions is acknowledged: Indian Institute of Chemical Biology, Calcutta; Indian Association for the Cultivation of Science, Calcutta; Jadavpur University, Calcutta; Central Glass and Ceramic Research Institute, Calcutta; Indian Toxicological Research Centre, Lucknow; Indian National Science Academy, New Delhi; World Health Organisation, New Delhi; Bhabha Atomic Research Centre, Bombay; Tata Institute of Fundamental Research, Bombay;

Mercury in the Ecosystem

Indian Institute of Geomagnetism, Bombay; British Council Library, New Delhi; NASA Johnson Space Centre, Houston, Texas, USA; Massachusetts Institute of Technology, Boston, Massachusetts, USA; Imperial College of Science and Technology, London, UK.

Department of Geological Sciences
Jadavpur University
Calcutta 700032

S.Mitra
26th January, 1984

CHAPTER I
MERCURY IN THE UNIVERSE

INTRODUCTION

Mercury (atomic number: 80, atomic weight: 200.59) occupies a unique position among the elements because of its high volatility. A study of the distribution of mercury in the cosmos throws some light on our knowledge of cosmochemical processes in space. Some intriguing problems arise out of such studies e.g., the high mercury content of meteorites conflicts with all of the current models for the chemical fractionation of the elements. Data on the distribution of mercury in cosmic bodies have also been helpful in establishing the interior composition of our planet (Saukov, 1946, Urey & Craig, 1953). This chapter presents an overview of the information available on the nature of the distribution of mercury in the earth and other celestial bodies.

In the case of the data compiled by Suess and Urey (1956), the elemental abundances in the universe exhibit a gradual decrease with increasing atomic number Their data indicate the abundance of Hg to be 0.003 atoms per 10000 atoms of Si in the universe. The cosmic distribution of Hg in solar and celestial bodies is not well known. All of the elements which are less volatile than Hg have been retained in the cosmic proportions in meteorites and in the earth (Urey, 1954).

THE EARTH

The average concentration of Hg in the crustal rocks is about 0.007ppm (Vinogradov, 1959). However, the mantle has a higher content. Overall, sea-water contains an average of 0.0003ppm, while Japanese sea-water contains 0.002ppm Hg, of which 90% is fixed by organisms. The content increases with increasing depth of the water. The Hg content generally increases in going from oceanic to inland seas. The aquatic life-forms also follow the same sequence. Fish from the Caspian Sea contain nearly twice as much Hg (0.275mg/kg) in their muscle as do fish from the North and Barents Seas which, in turn, contain twice as much as do fish from the Atlantic Ocean (Patin, 1983). In sea-water, Hg is first fixed in micro-organisms and algae and is generally more concentrated in large aquatic animals. In the atmosphere, the average concentration is about 2×10^{-7} to 8×10^{-7} ppm (Butcher & Clarkson, 1972).

The concentrations of Hg in some igneous rocks in the crust, as reported by early workers, are (in ppm):

Granite (Stock & Cucuel, 1934)	0.058
Granite (Preuss, 1941)	0.01
Gabbro (Stock & Cucuel, 1934)	0.079
Gabbro (Preuss, 1941)	0.1
Basic effusives (Saukov, 1946)	0.09

Neutron activation analyses for Hg indicate a gradual decrease from acid to ultrabasic rocks: USGS standard G-2 granite (0.039ppm) to dunite (0.004ppm). Deep-seated rocks (Kimberlite, 0.20ppm) and their various inclusions (up to 1.23ppm) contain higher concentrations. In regions of Hg mineralisation, the value may reach between 1 and 20ppm. Mercury occurs mainly as epithermal deposits. Cinnabar (HgS) is known to be deposited in hot springs. In sedimentary rocks, Hg is concentrated in hydrolysates by adsorption. Some of the values are:

Composite of 36 Palaeozoic marine shales (Stock & Cucuel, 1934)	0.51ppm
Sandstone, average (Preuss, 1941)	0.1ppm
Limestone (Stock & Cucuel, 1934)	0.033ppm

The Hg concentrations in some soils, as analysed by Stock and Cucuel (1934) were:

Humic horizon of forest soils	0.03 - 0.08ppm
Forest soils	0.10 - 0.29ppm
Cultivated soils	0.03 - 0.07ppm
Clay soils	0.03 - 0.034ppm
Sand	0.001 - 0.029ppm

In rocks and soils, mercury occurs as various combinations of oxides, sulphides, halides, organic complexes and native metal. Upon heating, different compounds are released at different temperatures. Therefore, temperature peaks can indicate the concentration of various phases. A persistent 105°C peak may represent Hg dissolved in moisture adsorbed on the mineral grains.

Relatively large amounts of mercury are found to be accumulated in the atmosphere. These are contributed by volcanic eruptions, industry, pesticide sprays, natural evaporation, and coal smoke. On average, air, rain-water and spring-water contain about $2 \times 10^{-8} g/m^3$, 2×10^{-4}ppm and 2×10^{-5}ppm of Hg, respectively. The natural cycle of mercury on the Earth is shown in figures 1.1.1 and 1.1.2.

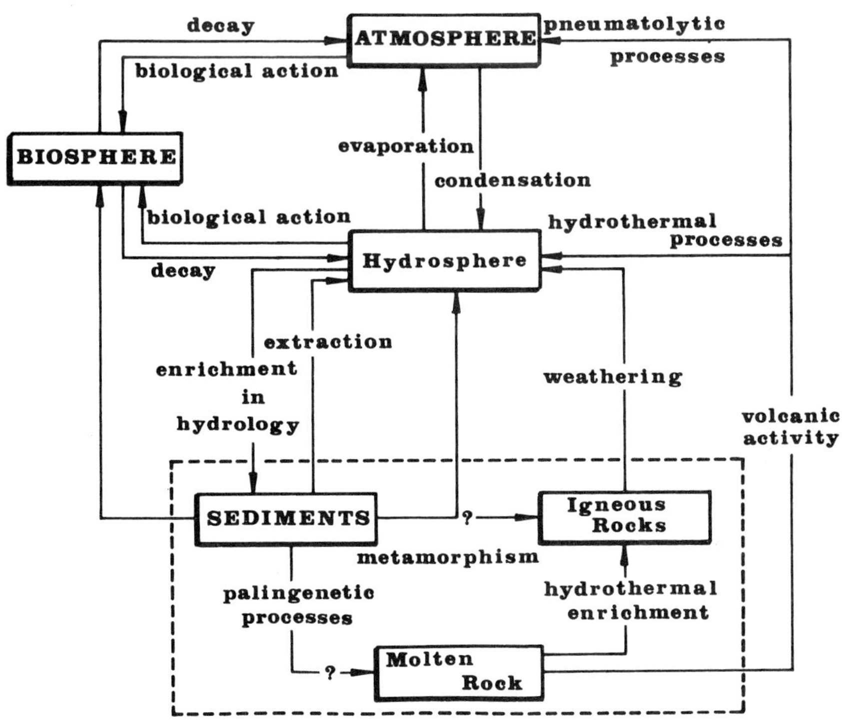

Figure 1.1.1 The Natural Cycle of Mercury on the Earth

Of the primary sources of mercury to the atmosphere, volcanic emanations are of prime importance. Atmospheric mercury levels measured at Kilaue and Mauna Loa volcanic systems in Hawaii and at Mount Erebus (Antarctica), Krysuvik, Myvatn, Geysir, and Hekla (Iceland) commonly show the same order of magnitude; amounting to between 10 and 25 $\mu g/m^3$ (Eshleman et al, 1971; Siegel & Siegel, 1973, 1978; Connor, 1979; Varekamp & Buseck, 1981). These values are several thousand times higher than the normal 'base-line' values discussed by McCarthy et al (1973).

Plants also influence the distribution of volcanic mercury by uptake from the atmosphere and from soil fluids, by transport to the shoot, and by re-release, as Hg°, to the atmosphere. Release rates as high as 2.5×10^{-4} g/h per kilogram have been observed.

On the other hand, the anthropogenic contribution to the atmosphere in continental areas has been calculated to be as high as 32% (Andren & Nriagu, 1979). This differs from the earlier Hg cycle proposed by Wollast et al (1975).

Table 1.1: Natural Geothermal Sources of Mercury (Siegel & Siegel, 1983)

Source	Mercury Content ($\mu g/g$)
(a) solids and liquids	
Mean crustal	50
Basaltic lava:	
USSR	6 - 1500
Iceland	67 - 815
Hawaii	20 - 1760
Antarctica (Mount Erebus)	0 - 3.6
Tuff (semiconsolidated fines)	
USSR	0 - 24000
Iceland (Surtsey)	7 - 85
Hawaii	350 - 375
Tephra (light ash)	
Antarctica (Mount Erebus)	0 - 3.6
Iceland (Surtsey)	350
Sublimates	
Hawaii	16 - 1700
Antarctic	1200 - 6250
Condensates	
USSR	0.2 - 72

Table 1.1 (continued): Natural Geothermal Sources of Mercury (Siegel & Siegel, 1983)

Source	Range	Mercury Content ($\mu g/g$) $\pm$ S.C.
	(b) gases ($\mu g/m^3$)	
Fumerolic		
Iceland	1.3 - 37.0	10.0 $\pm$ 5.2
Hawaii	1.0 - 40.7	17.5 $\pm$ 6.1
Volcanic		
Iceland	4.8 - 9.6	7.1 $\pm$ 0.7
Hawaii	0.7 - 200	29.2 $\pm$ 7.9
Antarctic (Mount Erebus)	3.6 - 24.1	11.3 $\pm$ 6.4
Non-thermal areas		
Iceland	0.62 - 1.0	0.8 $\pm$ 0.2
Hawaii	0.04 - 1.3	1.1 $\pm$ 0.5
California	$\leq$0.01	$\leq$0.01
New York	$\leq$0.014	$\leq$0.014
East Pacific	$\leq$0.001	$\leq$0.001

The natural geothermal sources of mercury, as compiled by Siegel and Siegel (1983) are presented in table 1.1.

The principal properties of mercury which are relevant to this monograph can be summarised for ready reference as:

Common form: Mercury globules, HgS (Cinnabar)

Associations: Sulphide minerals with Ag, Zn, Pb, Sb, Se, etc.

Aqueous species: Hg_2^{2+}, $Hg(OH)_2^{\circ}$, $HgCl_2^{\circ}$, Hg°, organomercurials

Average abundance:

Granitic rocks	0.04ppm
Basic rocks	0.01ppm
Ultramafic rocks	0.004ppm
Sandstone	0.03ppm
Shale	0.02 - 0.04ppm
Limestone	0.04ppm
Fresh water	0.07ppm

Abundances (continued):	Water supply ceiling	2ppm
	Air	$3ng/m^3$

Weathering products in soil: Hg, HgO, Hg_2Cl_2, $HgCl_2$, etc.

Applications as indicator element (a) in geochemical exploration: it serves as one of the best pathfinder elements for finding precious metals such as gold and silver, antimony, base metal sulphide deposits, and volcanogenic deposits; and (b) in environmental pollution: in tectonic and volcanically active zones and anthropogenically active (industrialised) zones.

METEORITES

In meteoritic matter, mercury is as abundant as lead (Eberhardt et al, 1970) and this is probably true throughout the solar system. Fleischer et al (1970) found evidence of fission tracks generated by nuclei with an atomic mass of 200. Half of the tracks could be due to mercury.

The Sun, via the solar wind and solar flares could be a significant source of Hg to the meteorites as well as to the moon. The cosmic abundance of mercury is 2.3 atoms per 10^6 silicon atoms.

Noddack and Noddack (1931) reported a Hg content of 200ppb in troilite from a Canyon Diablo meteorite. However, the same meteorite gave a value of 7ppb by later neutron activation analysis (Reed, Kigoshi & Turkevich, 1960). Ehman and Lovering (1967) have stated that calculation of the cosmic abundance of Hg on the basis of various chondrite models is made difficult by the extreme variability of the Hg content between different samples from a single meteorite, and between samples of different meteorites of the same general type.

The values of Hg abundance in various types of chondrite obtained by Ehman & Lovering (1967, pp357-376), together with values determined previously by Ehman & Huizenga (1959, pp125-135) and by Reed, Kigoshi & Turkevich (1960, p131) are reproduced in table I(1) in appendix I, and the values of the Hg abundances in various types of achondrite obtained by some investigators are reproduced in table I(2) in appendix I.

The accumulated data on the Hg contents of meteorites indicate that Hg is clearly fractionated in them and the order of abundance in meteorite types is: Carbonaceous -

1 ⩾ bronzite ⩾ carbonaceous - 2 > carbonaceous - 3 > Enstatite - 1 ≈ Enstatite - 2 ≈ amphoterite ≈ hypersthene ≈ bronzite.

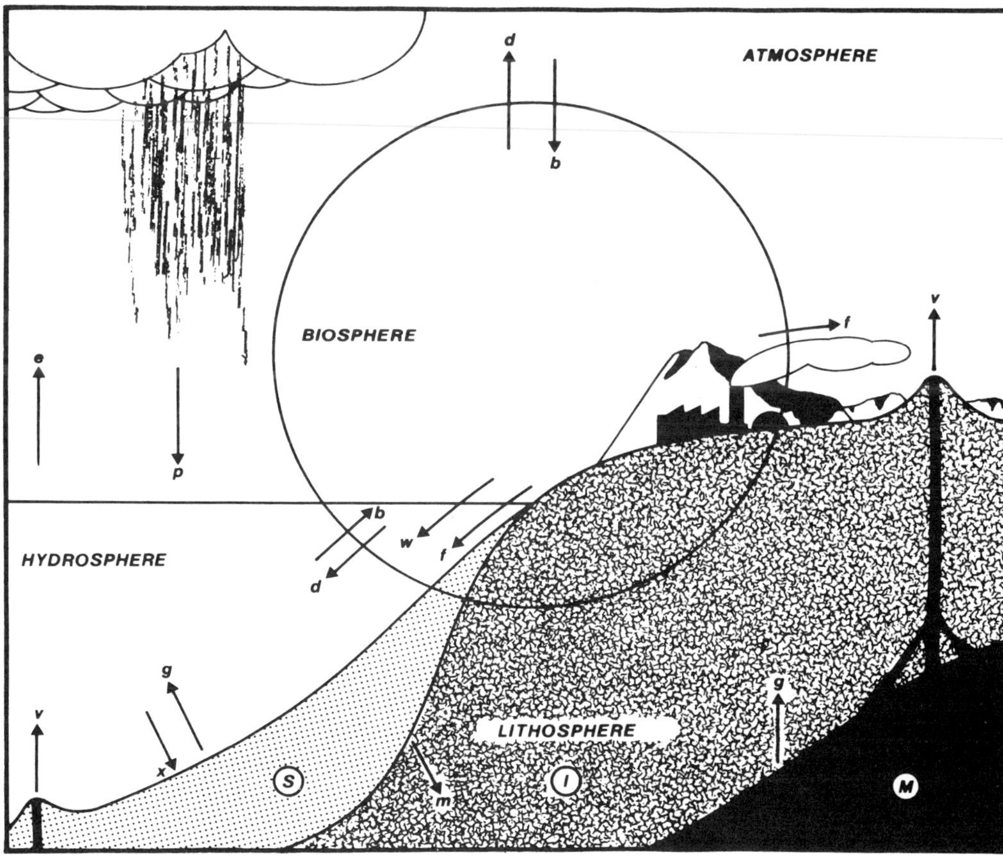

Figure 1.1.2 The Mercury Cycle. The cycle disperses the metal throughout the lithosphere, hydrosphere, and atmosphere. The biosphere encompasses all three. Mercury is present in all of the spheres in trace amounts, but tends to be concentrated by biological processes. Anthropogenic processes such as modern industry redistribute the metal to a marked extent and threaten the ecosystem. (I: igneous rocks, M: molten rocks, S: sediments, b: biological activity, d: decay, e: evaporation, f: discharge, g: enrichment, m: metamorphism?, p: precipitation, v: volcanic activity, w: weathering, x: extraction)

Reed, Kigoshi & Turkevich (1960) concluded that the much greater Hg concentration in the carbonaceous chondrite, Orgueil, compared to that in the enstatite chondrite, Abee, appears to be consistent with the thermal histories of the two meteorites. They have speculated that the carbonaceous chondrites must have been at a low temperature throughout their history, having retained quite considerable amounts of water and easily decomposed organic components.

Table 1.2: Characteristics of Mercury Distribution in Meteorites (Ozerova et al, 1973) - Mercury Content in Units of 10^{-8}g/g

No.	Class	n	$\bar{x}$	σ^2	S_k	σ_{S_k}	E_k	σ_{E_k}
1	Stony meteorites	186	664.3	4051.5	10.57	0.17	119.8	0.35
2	Chondrites without Carb	143	185.1	299.3	2.60	0.202	7.70	0.394
3	Olivine-bronzite	33	304.6	329.9	1.22	0.39	1.36	0.72
4	Olivine-hypersthene	96	206.5	525.2	5.843	0.243	41.02	0.472
5	Carbonaceous	24	3834.5*	10916	3.452	0.452	11.213	0.801
6	Achondrites	21	133.02	253.90	1.968	0.283	2.629	0.869
7	Feldspar-free	13	29.7	67.47	2.906	0.325	6.664	0.827
8	Feldspathic	8	300.91	352.63	0.614	6.201	1.294	0.725
9	Iron Meteorites	18	4.477	3.702	1.541	0.506	0.982	0.861
10	Meteorites of various classes containing troilite	10	24.36	45.107	1.868	0.612	1.913	0.691
11	Troilite from various classes of meteorite	6	241.5	471.40	1.605	0.481	0.452	0.423

n = number of analyses, $\bar{x}$ = mathematical expectation, taken to be the arithmetic mean, σ^2 = variance, S_k = asymmetry, σ_{S_k} = asymmetry error, E_k = excess, σ_{E_k} = excess error. * High average due to a uniquely high Hg content of the Orgueil meteorite

Ehman and Lovering (1967) felt that there are reasons for assuming that type I carbonaceous chondrites represent the closest approach to primitive solar system material.
The total Hg content of chondrites and achondrites, and the Hg released at temperatures greater than 450°C are presented in tables 1 to 5, appendix I.

Table 1.3: Comparison of Mercury Concentrations (10^{-8} g/g) in Troilite and Bulk Meteorites (Ozerova et al, 1973)

Class	Meteorite	Bulk Concentration	Troilite
Stony	Saratov	19	130
(chondrite)		65	-
(chondrite)	Tennasilm	140	1200
Stony-iron	Brahin	3.5	17
Iron	Sikhote-Alin	1.8	23
		1.8	-
		2	-
		2.8	-
		3.4	-
Iron	Santa Catharina	4.3	49

The nature of the distribution of Hg in meteorites is evident from tables 1.2 and 1.3, and from those in appendix I. On the whole, Hg is significantly more abundant in stony meteorites than in terrestrial rocks. The difference is even more striking if stony meteorites are compared with rocks of the upper mantle (from rifts in ocean zones, ophiolitic belts, kimberlite pipes, etc.) whose Hg abundance is lower than the clarke in rocks from the Earth's crust.
Stony and iron meteorites differ significantly in Hg content (table 1.3). Among the two types of stony meteorite, chondrite and achondrite, differences in Hg content are conspicuous. Among the two subgroups of chondrite, H and L, there is also an essential difference in Hg abundance. Carbonaceous chondrites have the largest Hg contents and also exhibit the largest variability, with values ranging over more than three orders of magnitude.

As already discussed, the high Hg concentrations in carbonaceous chondrites can be explained by processes analogous to the pneumatolytic and hydrothermal processes of terrestrial conditions. Evidence for such processes in carbonaceous chondrites has been obtained by a number of investigators, beginning with Zavaritsky (1948, 1950) and Kvasha (1948, 1963, 1968). Under terrestrial conditions, a Hg concentration as high as that found in these meteorites is found only in mercury deposits. In the achondrites, a significant difference is observed between the feldspathic and feldspar-free subgroups. The iron meteorites stand out from all other classes as having the lowest Hg concentration and the narrowest range.

Some workers (Ozerova et al, 1973) have attempted to specify the host minerals of Hg in meteorites. Troilite has been found to be a good concentrator (or rather host) for Hg with respect to the matrix material of the meteorites. This follows from the chalcophile property of troilite. A large part of the Hg in a meteorite sample is derived from its concentration in the troilites present. In iron meteorites, where the Hg concentration is usually low, the concentration in troilite would necessarily be low. On the other hand, in a Hg-rich meteorite, e.g., the Tennasilm chondrite, containing a higher bulk concentration of Hg (about 1ppm), the troilite phases contain 10ppm Hg (table 1.3). However, in some cases the higher concentration of Hg in troilite cannot explain the overall Hg content of meteorites. Certain samples of iron, stony-iron and stony meteorites (Santa Catharina, Sikhote-Alin, Brahin, Saratov & Tennasilm), have shown that the Hg content in troilite comprises less than 40% of the total present in the meteorites.

Studies of meteorites, using a Cameca MS-46 X-ray microanalyser and other methods, by Ozerova et al (1973) showed that neither the silicate part of the meteorites, nor the metallic iron, nor the troilite, contain any inclusions of Hg or Hg-bearing minerals. Therefore, they thought that Hg possibly occurs in an isomorphic form since up to 2% Hg can substitute isomorphously for Fe in pyrite.

RARE GAS - Hg CORRELATIONS

Reed and Jovanovic (1967) believe that Hg and the primordial gases became simultaneously associated in the chondritic materials. In these, Hg can be bound chemically or trapped in vacancies in the lattice.

It was previously thought that primordial gases are released near to the melting points of the silicates. However, Reed and Jovanovic (1967) showed that Hg is expelled at much lower temperatures. Thus, the Hg/Ar^{36}_{p} ratio can monitor the extent of subsequent heating. Three possibilities have been suggested: (1) low temperatures, Hg and Ar retained; (2) moderate temperatures (100 to 600°C), Hg partially lost and Ar retained; (3) high temperatures, both lost.

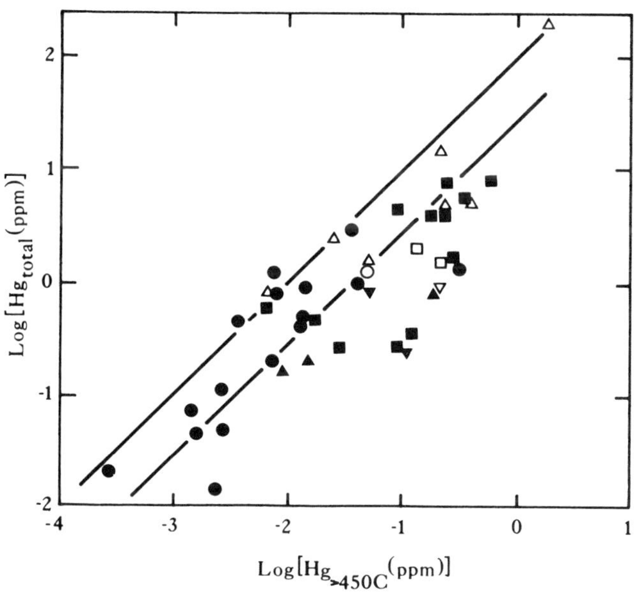

Figure 1.2 Total Mercury Content as a Function of the $Hg_{>450C}$ Content of Chondrites. (Closed circles: hypersthene chondrites, open circles: chondrules from hypersthene chondrites, closed squares: bronzite chondrites, open squares: chondrules from bronzite chondrites, diamonds: carbonaceous chondrites, unstable closed triangles: amphoterite chondrites, unstable open triangles: chondrules from amphoterite chondrites, stable closed triangles: enstatite chondrites).

Although there is no correlation between primordial Ar^{36} and ages, a parallelism exists between the Hg concentration and the K-Ar ages in the hypersthene chondrites

(Zahringer, 1964). Comparison of the ratio, Hg/Ar_p^{36}, with K-Ar ages reveals two distinct areas for hypersthene chondrites: a high Hg content being associated with long K-Ar ages and a low content being associated with short ages. Some exceptions are the Canakkale and Kryurka meteorites which have long ages but are among the lowest in Hg concentration, and Kunashak which has a short age but a high Hg content. It could be inferred from the above observations that some hypersthene chondrites were not heated, some were heated very early on and lost Hg, and others were heated late in their history and lost Hg and Ar^{40}.

Table 1.4: Mercury Concentration (ppb) in Apollo 11 Samples (Reed & Jovanovic, 1970)

Sample	Hg_{total}	$Hg_{>450C}$
Soil		
10084-1	0.95 ± 0.04	0.22 ± 0.02
10084-2	5.3 ± 0.1	0.06 ± 0.01
Microbreccia		
10061-22	1.2 ± 0.1	0.042 ± 0.002
Medium-grained rock		
10044-48	0.60 ± 0.03	0.028 ± 0.003
Vuggy rock		
10072-24	5.5 ± 0.1	1.1 ± 0.1
Vesicular rock		
	13 ± 2	0.09 ± 0.01

The enstatite chondrites show long K-Ar ages and low Hg concentrations and Hg/Ar_p^{36} ratios. These indicate early heating which was sufficient to remove Hg, relative to primordial Ar^{36}.

THERMAL RELEASE OF Hg

Reed & Jovanovic (1967) measured the release of Hg at temperatures of 110, 150, 200, 250, 300, 450, 600, and 1200°C, and observed that most of the meteorites fell into three release patterns. The major release of Hg usually occurred in a definite temperature interval. The optimum temperatures were between 150 and 250°C. At temperatures higher than this, a sharp drop in the Hg released was noticed.

Most of the Hg was released at temperatures below 450°C, where a minimum frequently occurred. This was followed by a small increase at higher temperatures. The release above 450°C is thought to be due to diffusion.

The correlation between Hg released at low and at high temperatures over a concentration range of 10^5 indicates that equilibrium was established in the distribution of Hg among the sites available in most meteorites.

THERMAL HISTORY

Measurements of Hg content provide evidence relevant to the thermal history of chondrites. Reed and Jovanovic (1967) isolated and analysed chondrules from pure meteorites. The relative amount of high temperature Hg in most cases is greater in the chondrules compared to that in the matrix of the meteorites. This suggests that complete equilibrium between the matrix and the chondrules was not attained. This lack of equilibrium could be accounted for by a low temperature history.

Meteorites having the largest relative amount of low temperature Hg are considered to indicate disequilibrium or no metamorphism, e.g., Orgueil, Karoonda and Kryarka meteorites.

The majority of the hypersthene chondrites are described as being equilibrated or metamorphosed. The metamorphosed ones have been at a higher temperature, and the curve of Hg_{total} versus $Hg_{>450}$ which best fits them is lower (higher relative amount of $Hg_{>450}$), figure 1.2. The general trend is a decrease in the ratio, $Hg_{total}/Hg_{>450}$, in going from (a) disequilibrium, unmetamorphosed, friable carbonaceous, and hypersthene chondrites to (b) metamorphosed bronzite and unmetamorphosed chondrites of all classes.

THE MOON

Lunar samples contain mercury. The day-time temperature of 130°C may cause volatilisation of mercury and continuous depletion through the tenuous lunar atmosphere. For mercury replenishment, the existence of an internal reservoir or exterior source (e.g. meteorite falls, solar wind) can be assumed. A striking difference between lunar rocks on the one hand and terrestrial rocks on the other hand is the frequent and large release of Hg (greater than 10%) at or below 130°C in the lunar

samples.
The results of analyses of samples from Apollo missions 11 and 12 are discussed below:

APOLLO 11
The Hg contents of some lunar samples obtained by Apollo 11 (Reed & Jovanovic, 1970) are presented in table 1.4. The values range from 0.6 to 13ppb. Vesicular rocks contain the largest values. However, these values are significantly lower than those in basaltic achondrites (18 - 9000ppb) or chondrites, and approach more nearly the values observed in some terrestrial rocks (Jovanovic & Reed, 1968).
The Hg release patterns from the lunar samples were significantly different to patterns observed in meteorites and terrestrial rocks (Jovanovic & Reed, 1968). A large release of 25 to 45% of the total Hg is observed at low temperatures (less than 110°C) and the maximum in the release is shifted from 150 to 250°C. During isothermal heating at 250°C, an initial burst occurs which involves 50 to 80% of the Hg. The large 110°C fraction is due to the release of free Hg adsorbed on the surface. It may be Hg released from low temperature (less than 150°C) traps during thermal cycling on the lunar surface, or Hg released during the degradation of surface material. A deviation from the trend was noted in the soil sample, 10084-1, which yielded a release pattern similar to meteorites and terrestrial rocks and which also gave the highest relative amount of Hg released above 450°C. Some of the low temperature Hg must constitute a tenuous atmosphere on the sunlit side of the moon. The fraction volatilised will depend upon the absorbancy of the surfaces.
Reed and Jovanovic (1970) interpreted the release trends as follows. The most degraded material, soil 10084-2, has the largest low temperature fraction, followed by microbreccia. The fine-grained rock, 10072-24, has a greater amount of its Hg in high temperature traps. The medium-grained igneous rock occupies an intermediate position, possibly because of slower cooling to a lower temperature. The relatively large low temperature release from the vesicular rock (10017-22) may be due to Hg which is associated with the vapour phase that formed the vesicles and which is trapped in the glass-like lining of the voids. Subtraction of the amount of Hg released below 110°C does not alter the trends noted.

APOLLO 12
The top, middle, and bottom of the 40cm core from Apollo 12 released 0.06, 0.3, and

16ppb of mercury, respectively. The total concentration and the concentration of mercury released at low temperatures from lunar samples (Reed et al, 1971) are presented in table 1.5.

Table 1.5: Mercury Concentration (ppb) in Apollo 12 Samples

No.	Type	Depth, Location	Hg at ≤130C (ppb)	Total Hg (ppb)
12025,74	core	1.5 - 2.5cm	0.06	10
12028,215	core	18 - 19cm	0.30	17
12028,216	core	38 - 39cm	16	20
12052,18	rock	interior	0.40	3.0
12052,49	rock	exterior	1.60	9.0
12053,31	rock	exterior	0.72	3.8
12053,53	rock	exterior	4.70	12.0
12022,109	rock	interior	3.5	8.0
12022,69	rock	exterior	1.8	7.6
12070,61	soil	>150μm	0.15	0.9
12070,61	soil	<74μm	0.13	3.3
12033,18	soil	-	1.0	2.6

The surface-related Hg concentrations cluster into two groups: 0.1 to 0.5ppb (predominantly soils) and 1 to 5ppb (predominantly crystalline rocks). The distributions of mercury in rocks and voids are determined by diffusion through voids, along surfaces associated with grain boundaries, cracks, fissures, and dislocations, and by volume diffusion through crystal lattices.

The maximum surface-related mercury concentration has an average value of about 3.4×10^{-9}g/g (= 3.4×10^{-9}g per $0.6cm^3$) of soil. This amount can be provided by a few tenths of a percent of meteoritic material with Hg abundances of 2.3 atoms per 10^6 Si atoms. In meteoritic matter, Hg is about as abundant as Pb (Eberhardt et al, 1970) and this is probably true throughout the solar system. A solar wind flux of 10^8protons/cm^2s will provide the 10^{11} atoms or 10^{-11}g/cm^2 of mercury which are estimated to be lost from the lunar surface during one million years (Reed et al, 1971).

In the lunar core studies, it was suggested by them that mercury becomes trapped in cooler interfaces between rock and soil (regolith) and, consequently, the study of surface-related Hg may help to establish the orientation of a lunar rock (Reed et al, 1971).

VENUS (atmosphere)

Recently, Lewis and Fegley (1982) have shown that the terrestrial crustal abundance of mercury would suffice to produce substantial masses of halide cloud condensates on Venus. Mercury is so volatile at the surface temperature of Venus that it would reside almost completely in the atmosphere.

The failure, of the 1978 Pioneer Venus mission, to detect even a trace of mercury in the lower atmosphere (Hoffman et al, 1979, 1980) strongly suggests that Venus is deficient in mercury, relative to the Earth. The evidence of severe mercury depletion is more consistent with a volatile-poor high-temperature origin for Venus.

Water and mercury (and perhaps arsenic as well) are depleted on Venus as compared with the Earth.

REFERENCES

- Andren, A. and Nriagu, J. (1979) in - The Biogeochemistry of Mercury in the Environment, (Edited by J.Nriagu), North-Holland Biomedical Press, pp 1-15

- Butcher, S.S. and Clarkson, R.J. (1972) - An Introduction to Air Chemistry, Academic Press, New York - London, p241

- Connor, J. (1979) - Science of the Total Environment, 12, 241-50

- Eberhardt, P. et al (1970) - Geochimica et Cosmochimica Acta, 2 (Supplement 1), 1037

- Ehman, W.D. and Huizenga, J. (1959) - Geochimica et Cosmochimica Acta, 20, 122

- Ehman, W.D. and Lovering, J.T. (1967) - Geochimica et Cosmochimica Acta, 31, 357-76

- Eshleman, A., Siegel, S., and Siegel, B. (1971) - Nature, 233, 471-2

- Fleischer, R.L., Haines, E.L., Hanneman, R.E., Hart, H.R., Kasper, J.S., Lifshin, E., Woods, R.T., Price, P.B. (1970) - Science, 167, 548

- Furzov, V.Z. (1973) - American Geological Institute, Doklady Academy of Sciences, USSR, 204, 184-7

- Hoffman, J.H., Hodges, R.R., Donahue, T.M., and McElroy, M.B. (1980) - Journal of Geophysical Research, 85, 7882

- Hoffman, J.H., Hodges, R.R., McElroy, M.B., Donahue, T.M., and Kolpin, M. (1979) - Science, 205, 49

- Jovanovic, S. and Reed, G.W. (1968) - Geochimica et Cosmochimica Acta, 32, 341-6

- Kvasha, L.G. (1948) - Meteoritika, 4, 83-96 (in Russian)

- Kvasha, L.G. (1950) - Meteoritika, 8, 116-8 (in Russian)

- Kvasha, L.G. (1963) in - Origin of Life in the Universe 44-55, Academy of Sciences of the USSR (in Russian)

- Kvasha, L.G. (1968) in - Abiogenesis and First Stage in Evolution of Life 16-22, Nauka (in Russian)

- Lewis, J.S. and Fegley, B. (1982) - Science, 216, 1223-5

- McCarthy, J., Meuschke, L., Ficklin, W., and Learned, R. (1973) - US Geological Survey Professional Paper, 713, 37-41

- Noddack, I and Noddack, W. (1931) - Zeitschrift für Physikalische Chemie, 154, 223-44

- Ozerova, N.A, Kvasha, L.G., Bulkin, G.A., and Aidinian, N.K. (1973) - Geochimica et Cosmochimica Acta, 37, 569-82

- Patin, S.A. (1983) - Pollution and Biological Resources of the Oceans, Butterworth Scientific, London, p289

- Preuss, E. (1941) - Angewandte Mineralogie, 3, 6

- Reed, G.W. (1971) - Mercury (80) in Handbook of Elemental Abundances in Meteorites (Edited by B.Mason), Gordon & Breach, 487-91

- Reed, G.W. Goleb, J.A., and Jovanovic, S. (1971) - Science, 172, 258-61

- Reed, G.W. and Jovanovic, S. (1967) - Journal of Geophysical Research, 72, 2219

- Reed, G.W. and Jovanovic, S. (1968) in - Origin and Distribution of the Elements (Edited by L.H.Ahrens), Pergamon Press, Oxford, 323-8

- Reed, G.W. and Jovanovic, S. (1970) - Proceedings of the Apollo 11 Lunar Science Conference, Vol.2, Pergamon Press, New York

- Reed, G.W., Jovanovic, S., and Fuchs, L.H. (1970) - Science, 167, 501
- Reed, G.W., Kigoshi, K., and Turkevich, A. (1960) - Geochimica et Cosmochimica Acta, 20, 122
- Saukov, A.A. (1946) - Trudu Instituta Geologicaeskogo, Akademii Nauk SSSR, 78(17), 129
- Siegel, B.Z. and Siegel, S.M. (1973) - Nature, 241, 526-7
- Siegel, B.Z. and Siegel, S.M. (1978) - Environmental Science and Technology, 12, 1036-9
- Siegel, B.Z. and Siegel, S.M. (1983) - Advances in Space Research, 3(9), 135-9
- Stock, A. and Cucuel, F. (1934) - Naturwissenschaften, 22, 390-3
- Suess, H.E. and Urey, H.C. (1956) - Reviews of Modern Physics, 28, 53-74
- Urey, H.C. (1954) - Astrophysics, 1, 147
- Urey, H.C. and Craig, H. (1953) - Geochimica et Cosmochimica Acta, 4, 36-82
- Varekamp, J. and Buseck, P. (1981) - Nature, 293, 555
- Vinogradov, A.P. (1959) - The Geochemistry of Rare and Dispersed Chemical Elements in Soils, Chapman & Hall, London, 209pp
- Vinogradov, A.P. (1962) - Geokhimiya, 7, 555-71 (in Russian)
- Wollast, R., Billen, G., and Mackenzie, F. (1975) in - Ecological Toxicology Research: Effects of Heavy Metal and Organohalogen Compounds, Proceedings of a NATO Science Conference, pp145-66
- Zahringer, J. (1964) - Meteoritika, 22, 25-40
- Zavaritsky, A.N. (1948) - Papers of the Mineralogical Society of the USSR Ser, 2, 77, 117-24 (in Russian)
- Zavaritsky, A.N. (1950) - Meteoritika, 8, 100-15 (in Russian)

CHAPTER II

MINERALOGY AND GEOCHEMISTRY OF MERCURY

INTRODUCTION

Geochemically speaking, Hg has a chalcophile character. This chalcophile nature is, according to Fersman, indicated by its position among the elements of the sulphide deposits and by its atomic volume curve. At magmatic temperatures, according to Goldschmidt (1954), the redox potential of ferrous iron in magmas is, in most cases, sufficient to transform Hg compounds into metallic Hg and thus force the metal into gaseous exhalation products, where it can again combine with sulphur at a later stage.

MINERALOGY

Mercury forms a large number of compounds, and the major Hg-bearing minerals are presented in table 2.1.

Among these, the most common are native Hg and cinnabar. The latter contains 86.2% Hg and 13.8% S, and belongs to the trigonal-trapezohedral crystal system ($P3_121$, a = 0.4146nm, c = 0.9497nm, Z = 3, a_{rh} = 0.3969nm, α = 62.58. The structure of cinnabar is of distorted NaCl (PbS) type. Each Hg atom is surrounded by six S atoms at the corners of a distorted octahedron. The screw arrangement of the atoms,

-S-Hg-S-Hg-S-Hg-S-Hg-S

along the principal c-axis is a noteworthy feature. In the crustal rocks, the average content of Hg is only 0.02ppm (Taylor, 1964; Wedepohl, 1978).

OCCURRENCE

Major Hg deposits occur along two Quaternary - Tertiary orogenic-cum-volcanic belts

Table 2.1: Mercury Minerals

Name	Composition	Crystal Symmetry
Amalgam	(Hg,Ag,Au)	cubic
Arquerite	$HgAg_x$	cubic
Calomel	Hg_2Cl_2	tetragonal
Cinnabar	HgS	rhombohedral
Coccinite	HgI_2	tetragonal
Coloradoite	HgTe	cubic
Corderoite	$Hg_3S_2Cl_2$	-
Eglestonite	Hg_4OCl_2	cubic
Gaudalazarite	(Hg,Zn)S	cubic
Kleinite	$Hg_2N(Cl,SO_4),xH_2O$	hexagonal
Konigsbergite	HgAg	cubic
Leviglianite	(Hg,Zn,Fe)S	cubic
Livingstonite	$HgSb_4S_7$	monoclinic
Magnolite	Hg_2TeO_4	-
Metacinnabar	$HgS\text{-}(HgS)_{80}(HgSe)_{20}$	cubic
Montroydite	HgO	orthorhombic
Moschellandsbergite	Hg_3Ag_x	cubic
Mosesite	$Hg_6(NH_3)_2Cl_2(SO_4)(OH)_4$	cubic
Onofrite	$(HgS)_{20-80}(HgSe)_{80-20}$	cubic
Potarite	Hg_2Pd_3	cubic
Schwarzite	$(Cu,Fe,Hg)_{12}Sb_4S_{13}$	cubic
Schuetteite	$HgSO_4,2H_2O$	-
Terlinguaite	Hg_2OCl	monoclinic
Tiemannite	HgSe	cubic

(figure 2.1): (a) the circum-Pacific and (b) the Himalayan - Mediterranean. The deposits may be epigenetic or syngenetic.

Mercury is also commonly associated with epithermal Pre-Cambrian Zn-Pb-Cu deposits which exhibit very high contents of Hg whereas, in some syngenetic sulphide deposits, the Hg contents are exceptionally low, as in Kupferschiefer type and

Mississippi valley type ore deposits.

Saukov (1946) showed that the Hg contents of the following deposits increase systematically with decreasing intensity of genetic conditions: magmatic rock (77ppb), pegmatites (n ppb to 10n ppb), pneumatolytic deposits (10n ppb to 100n ppb); mesothermal deposits (100n ppb to n ppm); and epithermal deposits (n ppm to n%). This has been corroborated by later workers (Ozerova, 1962).

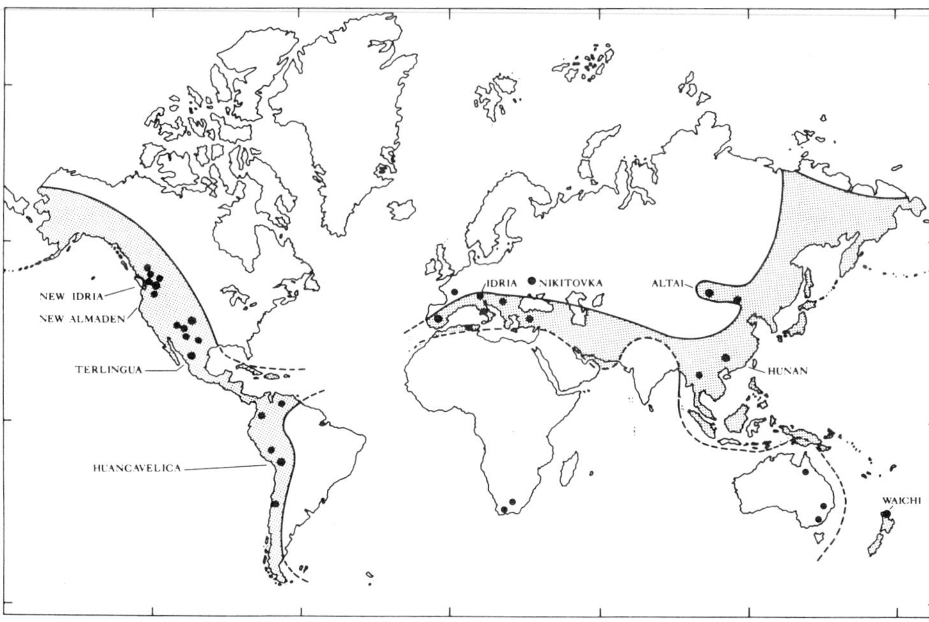

Figure 2.1 World Distribution of Important Mercury Deposits

The distribution of Hg in different rock types and soil has been presented in US Geological Survey Professional Paper No. 713, 1970.

The average and range of variation of the Hg content in several hundred common rocks from Sweden, reported by Ferm and Larsson (1973) are (in ng/g):

granite	3.5 (1.4 - 281)
diorite	5.0 (0.4 - 19.7)
diabase	3.9 (0.2 - 17.7)
sandstone	2.5 (0.8 - 6.0)
shales	5.9 (0.9 - 33.5)
bituminous shales	234 (31.9 - 340)
limestones	6.0 (0.8 - 31.2)

Normally, shales have higher contents of Hg, particularly when organic matter is present.

Most of the granitic rocks have contents of less than 2000ppb Hg. Two types of igneous rock: deep-seated eclogites and Kimberlites and alkalic rocks, exhibit higher contents of Hg, with averages ranging from hundreds to thousands of ppb of Hg. Analyses of the individual minerals of the alkalic rocks reveal a fairly uniform distribution of Hg in the main rock-forming minerals, and higher concentrations in some of the accessory minerals of high calcium, strontium, and barium content. All types of rock from such areas as the Donets Basin and Crimea in the USSR show very high Hg contents. The reason could lie in their proximity to areas of Hg mineralisation and mud volcanoes.

A statistical analysis by factor analysis and multiple regression of Hg, Cu, and other elements of Lake Dufault granodiorite and Rougemont intrusive, Quebec, suggests the existence of a strong association of Hg with Cu and S. Here, the presence of Hg cannot involve vapour transport since Cu is not likely to be transported by vapour (Webber & Newbury, 1971).

SULPHIDE DEPOSIT TYPES

Because of its chalcophile character, Hg is commonly associated with epithermal base-metal sulphides as well as with noble metal deposits. Volcanogenic sulphides, e.g. Zn-Pb-Cu deposits, are often of very high Hg content.

Sulphides of volcanogenic origin are generally found to have higher Hg levels than those of sedimentary strata-bound Kupferschiefer type and Mississippi Valley type deposits. Of the volcanogenic sulphides, higher Hg levels are found in the deposits of Proterozoic age compared to those in Archean or Phanerozoic times.

Vein or replacement deposits of Cu, Pb, and Zn in sedimentary rocks, apparently transected by deep faulting in some cases, are found to have high Hg contents. This type covers the Pre-Cambrian (Grenvillian) deposits of Balmat, New York (1100ppm), New Calumet, Quebec (53ppm) and paleozoic deposits in Ireland such as Gortdrum (economic Hg content), Keel (227ppm; Watling et al, 1973) and Tynagh (18ppm; Donovan & James, 1967).

Early Phanerozoic Fe-Zn-Cu deposits with modestly anomalous Hg levels (1ppm) are found in the middle Devonian Chalcopyrite-sphalerite deposits of West Shasta, California (Friedrich & Hawkes, 1966). Very much higher Hg levels are found in deposits which evolved later in the orogenic cycle such as those found in the

Pb-Zn-Cu-Ag deposits, viz., Mount Isa, Queensland, Kuroko deposits, Rosebury, Tasmania. High levels in the soil gas anomalies are found at some deposits of this class, e.g. Bad Grund and Rammelsberg, West Germany, Captains Flat, New South Wales, and Tharsis, Spain.

TECTONIC SETTING

(a) Plate margins and volcanic belts: The consuming plate margins represented by extensive linear chains of Mesozoic-Cenozoic tectonic activity contain most of the world's known Hg resources. In these zones, oceanic crustal material is destroyed by subduction and subsequent melting under the continental crust. At the consuming margin, crustal thickening occurs and deep-rooted volcanic systems act as conduits to transport and concentrate the Hg which is "cooked out" of the descending slab of overlying crustal material heated by the rising magma. Most Hg deposits occur on the continental side of post-Jurassic subduction zones (Bailey, 1975). Deep abyssal faults (or perhaps Paleozoic Benioff zones) that have been reactivated may be responsible for localising Hg deposits, viz., in southern Siberia (Kuznetsov et al, 1974).

Mantle processes are operative in oceanic ridge systems where subcrustal basic magmas are continuously accreted to the lithospheric plates. Such mantle processes are manifested by "hot spots" such as the islands of Hawaii and the associated chains of sea-mounds. High atmospheric concentrations ($20 mg/m^3$) have been observed near to volcanic activity in Hawaii. Similarly, high levels of atmospheric Hg ($1.1 \times 10^{-7} g/m^3$) i.e. a factor of 10^4 above normal atmospheric level, have been recorded on the volcanic islands close to Iceland, which straddle the Mid-Atlantic ridge (Siegel et al, 1973; Eshleman et al, 1971). Sea-water samples collected at depth near to the Mid-Atlantic ridge were also found to have markedly anomalous Hg contents (80 - 400ng/l) compared to the normal values of 2 - 40ng/l (Carr et al, 1975).

As shown in figures 2.2.1 to 2.2.5, the distribution of Hg deposits is mostly confined to the tectonically active subduction and volcanic zones at the consuming plate margins. Some of the Hg mineralisation is zoned along the ophiolite (with serpentinite, etc.) or mantle diapirs and outer-arc marginal basins, where Cyprus-type Cu-Fe deposits are commonly found (Mitchell & Garson, 1976, p131) (figure 2.2.1). The other major type of Hg mineralization occurs in the oceanic rise, Hawaii-type (figure 2.2.2) and Andean-type magmatic belt (figure 2.2.3). A schematic cross-section depicting mineralisation in an island arc with a well-developed outer arc is shown in figure 2.2.4

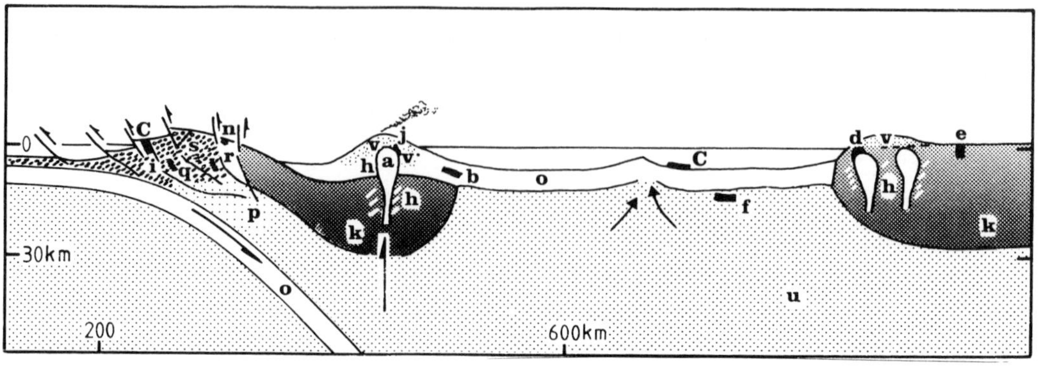

Figure 2.2.1 Mercury Deposits in Magmatic and Outer Arc with Marginal Basin. (a: tonalite granodiorite, b: Besshi-type Cu-Fe, C: Cyprus-type Cu-Fe, d: Sn, W, Bi, Mo, F, e: Sb, f: podiform Cr, NiS, Pt, h: high-temperature metamorphics, i: serpentinite, j: porphyry Cu-Au, k: continental crust, n: podiform Cr, o: oceanic crust, p: high-pressure metamorphics, q: quartz veins, r: ophiolite or mantle diapir, s: sediments, u: upper mantle, v: volcanic rocks [calc alkaline rhyolite], v': volcanic rocks [island arc tholeiite])

(Mitchell & Bell, 1973).

Towards the continental side of the arc-trench are areas of rifting, faulting, plutonic activity, and waning volcanism where hydrothermal Au, Cu, and Hg mineralisation takes place. Further away from the margin, porphyry-type Cu, Au, and Hg deposits form, with the development of calderas and granodiorite plutons in volcanic arcs. The porphyry Cu deposits occur below the zone of Hg deposition. The interrelationships of the structural and tectonic features in such deposits are shown in figure 2.2.5.

Mercury deposits occur in active or slightly eroded volcanoes within some of the more complex modern island arcs, e.g., the Phillipines, Japan, and New Zealand. White (1974) studied porphyry Cu deposits and opined that Hg could be deposited in vapour-dominated reservoirs above boiling brine zones in which porphyry Cu ores are found.

The mercury contents of andesites and basalts of continental origin are found to be higher than those of similar samples originating from island arcs or the Circum-Pacific belt (Flanagan et al, 1982).

<u>b) Lineaments:</u> Besides mineralised zones, Hg 'anomalies' are found to be localised

along some lineaments in the crust of the earth. These anomalies are due to the process of degassing along deep fault zones during their activation, the sources being located at mantle depths, rather than within the crust. High Hg values in some stony meteorites and kimberlites indicate that the mantle is possibly richer in Hg than are the crustal rocks. The east-west "48th Parallel Lineament", with faults and intrusions, extends 800 miles from northern Virginia to south central Missouri (Heyl, 1972). In this zone, extremely anomalous Hg values have been recorded from some ores of the Central Kentucky and Illinois Kentucky districts close to this lineament (Heyl, 1970).

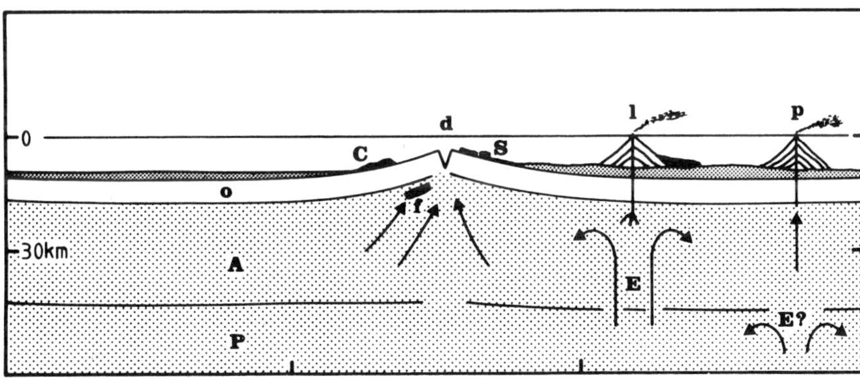

Figure 2.2.2 Mercury Mineralisation Occurring in the Oceanic Rise and Hawaii-Type Chain. (A: alkali-depleted mantle, C: Cyprus-type Cu-Fe, d: LIL-depleted basalt, E: plume, f: podiform Cr, NiS, Pt, l: LIL-rich basalt, o: oceanic crust, p: potassic basalt, P: primitive mantle, S: As, B, Fe, Hg, Mn, Pb, Zn sediments)

<u>c) Riftzones:</u> These normally do not bear significant amounts of Hg. No occurrence of Hg has been reported from the East Africa, Red Sea rift system (Jonasson & Boyle, 1971). If it is taken to be due to a lack of data, the reported amount of Hg in the Red Sea deposits is less than 10ppm in some mineralised cores (Degens & Ross, 1969).

The Baikal riftzone in south-eastern Siberia is a major rift system which is some 1500km in length and developed mainly during the late Pliocene-Pleistocene. However, all of the reported Hg mineralisations in the Baikal Belt are Permo-Carboniferous in age.

Similarly, in the landward extension of the embryonic rift presented by the Gulf of California, no significant occurrence of Hg minerals has been reported from the geothermal areas of Imperial Valley or Salton Sea. However, 8ppb of Hg was reported only from drill samples of metalliferous brines encountered at depth in the Salton Sea (White, 1967).

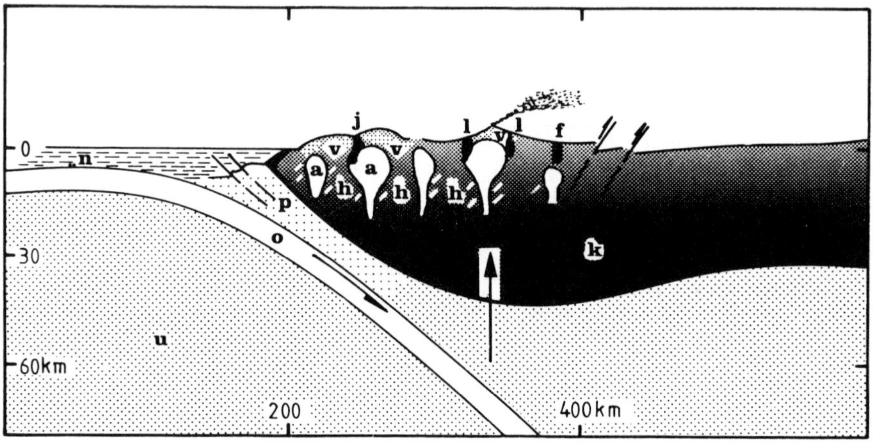

Figure 2.2.3 Mercury Deposits in the Andean-Type Magmatic Belt. (a: tonalite granodiorite, f: F, h: high-temperature metamorphics, j: porphyry Cu, Mo, Hg, k: continental crust, l: Sn, W, n: Mn nodules with Cu, Ni, Co, o: oceanic crust, p: high-pressure metamorphics, u: upper mantle, v: volcanic rocks)

ROCK ASSOCIATION

Kuznetsov and Obolenskii (1969) described the genetic association of Hg ores with derivatives of basaltic volcanism in areas of recent volcanic activity; and their relation in time with dyke rock complexes in folded regions, areas of tectonomagmatic activation of fold margins and also of the central massif and ancient platforms. Both igneous intrusions and Hg mineralisations are the derivatives of deep-seated subcrustal basaltic magma.

Stakeev et al (1975) presented results of the fluorescence spectrometric estimation of Hg in basalts, andesites and rhyolites from Iceland; trachytes, basalts, melilites, carbonatites and phonolites from the East African Rift zone; alkali felspathoidal rocks from Guinea; alkali rocks from Greenland, and uralites, foyaites, from Lovozersk massif (USSR). Like many other terrestrial elements, Hg exhibits a log normal distribution pattern in geological samples.

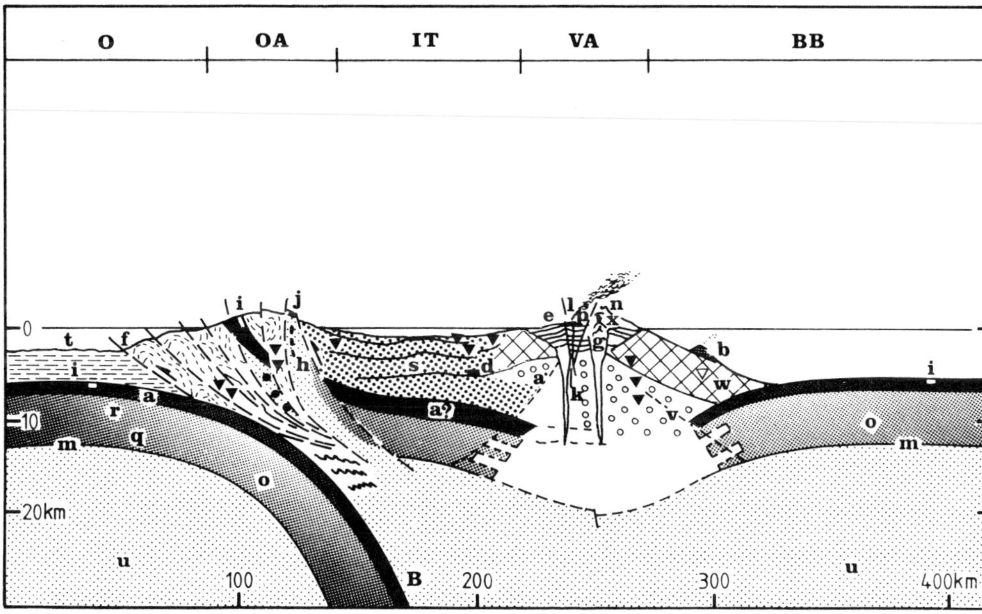

Figure 2.2.4 Schematic Cross-Section Illustrating Mineralisation in an Island Arc with Well-Developed Outer Arc (Exaggerated vertical scale - Mitchell & Bell, 1973). (a: pillow lava, a': basaltic pillow lava [calc-alk], b: calc-alk or alk lava, B: Benioff zone, d: Besshi sulphides, e: sub-aerially erupted daciteandesite [calc-alk], f: subduction zone, g: porphyry Cu-Au, h: podiform chromite, i: Cyprus-type pyritic Cu, j: Ni laterite, k: melange and glaucophane schist, l: Kuroko Zn-Cu-Pb, m: moho, n: native sulphurpyrites, o: oceanic crust, p: dacite and tholeiite, q: gabbro harzburgite, r: dolerite dykes, s: marine and continental sediments, t: turbidites and mudstones, u: upper mantle, v: auriferous quartz veins, w: volcanoclastics [calc-alk], x: Hg, open circles: metamorphosed oceanic crust or continental crust remnants, closed circles: Hg in serpentinite, open triangle: high temperature pressure metamorphic boundary, closed triangles: placer Au, O: ocean, OA: outer arc, IT: inter-arc trough, VA: volcanic arc, BB: back-arc basin)

The Hg contents in the basaltic and alkali basaltic dykes and pipes of the Zeravshan-Hissar zone are 8×10^{-6} and $1.2 \times 10^{-5}\%$ respectively. Alkali basaltic rocks of the South Hissar zone contain $10^{-6}\%$. These are considered to be the host of primary syngenetic ores (Mogarovskii et al, 1975).

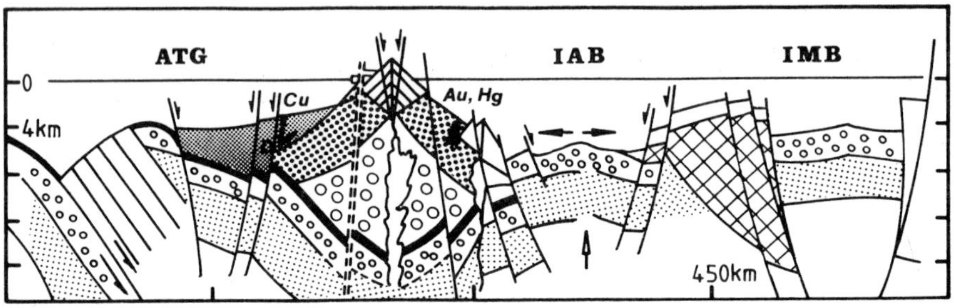

Figure 2.2.5 Gold-Copper-Mercury Deposits Associated with Tectonism, Acid Magmatism, Arc Rifting, Faulting, etc., in Arc-Trench to Marginal Basin Region. (Plutonic activity, waning volcanism, faulting, and arc rifting. Plutons in volcanic arc, caldera development with formation of porphyry Cu, Au, and Hg deposits, and sedimentation in arc-trench gap, development of inter-arc basin. Porphyry Cu occurs below Hg. ATG: arc-trench gap, IAB: inter-arc basin, IMB: inactive marginal basin)

Mercury in rocks and minerals of the Skaergaard intrusion, Greenland, were studied by Dissanayake and Vincent (1975). Using neutron activation analysis, they analysed nineteen representative rock samples and separated cumulus minerals from five rocks. The Hg values were 0.23ppm in chilled marginal gabbro and 0.07 - 0.34ppm in the layered series of rocks and reached 1.29ppm in one exceptional case. The Hg content is higher in leucocratic than in melanocratic rocks, even when occurring in the same horizon. Higher levels of Hg are found in the granophyric rocks at the top of the intrusion. The distribution of Hg in various cumulus phases is uniform and involves uncharged atoms occupying mainly interstitial spaces. High volatility and chemical inertness are found to be the major factors controlling the distribution pattern of Hg in the Skaergaard intrusion; no significant evidence of chalcophilic character has been observed. Mercury concentration leading to ore deposition in leucocratic differentiated rocks, viz., granite, has also been found in Uchursai (Burikhin & Burikhina, 1974).

Correlation and grouping of elements based on factor analysis and multiple regression analyses of the data of the contents of Hg, Cu, Zn, Fe, S, etc., in rocks and ore from New Calumet mine, Quebec and Labrador trough were done by Smith and Webber (1972).

The patterns of geographic distribution and formation of Hg deposits have been considered by many scientists but the genetic pattern of shallow epithermal deposits are as yet poorly studied. Investigations by Menyaylova (1967) on the relationship between Hg mineralisation and hypabassal diabase revealed the following:

1) The late magmatic alteration of diabase was affected by propylitisation accompanied by albitisation, chloritisation, calcitisation and lesser sericitisation and clinozoisitisation.
2) Propylitisation affected the entire diabase body, whereas the zones of intense alteration were of limited extent and were confined to zones of greater tectonic activity. The existence of such zones and the intrusive diastrophism is manisfested by numerous slickensides, xenoliths, and rheomorphic veinlets, found together with stringers mineralised by Hg. The diabase intrusion may have been emplaced in a large tectonic fissure acting simultaneously as a channel for mineralising solutions.
3) Hydrothermal alteration of rocks has led to replacement of early minerals. The rocks formed consist of ferromagnesian carbonates, dickite, and quartz containing disseminated cinnabar.
4) Many diabase porphyry dykes contain finely disseminated cinnabar.

Propylites with Hg-As (cinnabar-realgar) and Hg (cinnabar) mineralisation occur in intermediate to acidic intrusives and extrusives in volcanogenic and terrigenous beds (Uzgalis, 1975). Gold-bearing propylites are localised in basic to intermediate sub-intrusive bodies in essentially volcanic rocks. Propylites and secondary quartzites with Sb-Hg (stibnite-cinnabar) mineralisation are localised along faults and have no direct connection with magmatic bodies. The degree of Hg and Au dissemination decreased from early to late stages of regional hydrothermal metasomatism. Ore-bearing quartz in metasomatites with Hg mineralisation formed at about 220°C but, in Au-bearing sulphides, it formed at 280 to 400°C. As in Kamchatka, the south-eastern Koryak range has vertical ore zoning.

Many Au ores have an extremely low content of Hg (Sears, 1971) whereas some deposits, including the Dadsen (Red Lake, Ontario) and the Ross mine (Ramore, Ontario) contain much higher levels. Tennantite had been observed at the Ross mine and Coloradoite is reported from Porcupine and Kirkland Lake ores. If the original deposition is volcanic exhalative, and introducing cinnabar as ore of the Hg minerals, its presence in remobilised mineralisation depends on sulphides and sulpho-salts present during high temperature events. Varying amounts of cinnabar associated with the original Au mineralisation could be a significant factor in accounting for the present Hg content.

A high level of Hg concentration is also found in association with other volcanic exhalative deposits, such as W-Sb-Hg mineralisation belonging to the Ordovician and Silurian ages. A large part of the younger W deposits associated with granites derived their W content from the older strata-bound scheelite deposits during granitisation. Such mineralisation is reported from Sangdong (Korea), Felbertal (Austria), and Orsdalen (Norway). One stratiform scheelite deposit in the Austrian Alps was reported to show a strong Hg soil gas anomaly. Mercury occurs in the original bedded deposits, primarily as cinnabar, and in the absence of suitable 'trap' sulphides would be mobile during metamorphism and be dispersed as a weak aureole in the host rocks of any subsequent mineralisation.

The (synthetic) heavy metal sulphides are relatively unstable in aerobic soils; the degree of oxidation more or less increasing with increase in the solubility of the sulphides. Under anaerobic soil conditions, the metal sulphides remain stable.

In the massive sulphide deposits of Japan, the Kuril islands, and the Urals, Hg is lower in metamorphosed than in non-metamorphosed ores. Two varieties of Cainozoic sulphide ores can be distinguished: 1 - products of sub-continental solfataric action with high Hg in pyrite with cinnabar, 2 - products of post-volcanic activity in submarine suites with low Hg in sulphides (Kuroko type). In co-existing sulphides: Hg in sphalerite $>$ Hg in chalcopyrite $>$ Hg in pyrite; but some sulphosalts as well as gold and tellurides concentrate Hg as sphalerites. During metamorphism, sulphides lose Hg which leads to the formation of Hg ores at some distance from metamorphosed sulphide bodies in the same region.

DEPOSITS IN RELATION TO HYDROTHERMAL PROCESS

White's (1974) study has indicated that five types of water (meteoric, ocean, evolved connate, metamorphic, and magmatic) account in varying proportions for hydrothermal fluids. The regions studied include: 1 - Hg systems involving metamorphic or evolved connate waters, 2 - epithermal Au-Ag districts and similar active hydrothermal systems involving meteoric water and perhaps as much as 5 or 10% of magmatic water, 3 - ore districts and active systems with abundant base metal, involving waters of diverse origin: ocean, evolved connate, meteoric and magmatic, usually with salinites, 4%, and 4 - Cu-Mo deposits, generally involving magmatic water but with proportions of meteoric waters during sulphide and post-sulphide stages. Individual constituents, including metals, Cl, S, and CO_2, may come from the same source as the dominant water, or from different sources.

In serpentinites, mercury ores (mainly cinnabar) occur principally in listownites. The serpentinites formed from autometamorphism of dunite and wehrlite and are hydrothermally altered to listownites in three stages: 1 - a high temperature stage of carbonisation accompanied by albitisation, 2 - development of carbonate and opal-containing rocks with albite and 3 - formation of breunnerite and chalcedony containing rocks with zeolites. During hydrothermal alteration of dunite and wehrlite, the relative mobility sequence of components is H_2O, CO_2, K_2O, Na_2O, CaO, FeO, Al_2O_3, SiO_2, MgO, and TiO_2. Other rocks occurring in the Hg ore region are usually hydrothermally altered: the low temperature argillisation of the associated sedimentary rocks is effected by replacement with layered alumino-silicates (montmorillonites) while basalt porphyrites, tuffs and gabbro-diabases are altered to greenstones (Shashkina et al, 1972).

Ushak and Trufanov (1968) discussed the migration capacity of mercury sulphide and alumina under telethermal conditions. The telethermal quartz-cinnabar-dickite ore deposit in the north-western Caucasus is localised in Lower cretaceous terrigenous sedimentary rocks. The low temperature telethermal ore-bearing solutions were enriched in CO_2 and H_2S at 120 to 170°C. Experiments performed in an autoclave indicate that Al_2O_3, carbonates and HgS, under telethermal conditions, migrate in solutions of low pH to deposit crystals of cinnabar, dickite, and aragonite. Mercury migrates predominantly as $Na_2[Hg(CO_3)_2]$ or as $Hg(HCO_3)_2$ carbonate complexes. Alumina is transported as a complex compound of the form, $Na_x[HgAl_nO_m]$.

Baev (1968, 1974) studied the Hg in the supergene zones of the north-western Caucasus where the main ore deposit is cinnabar, which is extremely stable chemically in the supergene zone and migrates mechanically to form placer deposits. Mercury also migrates in the supergene zone in the sorbed, gaseous, and dissolved forms (Saukov et al, 1972). Redistribution of Hg occurs not only in the ore deposits but also in the endogenic dispersion haloes. Waters in the dispersion halo have a low H^+ concentration. The Hg content in the haloes is several times greater than its background value. The possibility of Hg transfer as sulphate and chloride complexes is indicated by a positive correlation between Hg and SO_4^{2-} and Cl^- ions.

A functional linear relationship exists between the solute content of HgS and the concentrations of Na_2S and hydroxyl complexes including $Hg(OH)_3^-$, $HgS(HS)(H_2O)_2^-$, $Hg(HS)_2(OH)_2^-$, $Hg(HS)_2(OH)_2^{2-}$, etc. Based upon the solute data of cinnabar and metacinnabarite, the thermodynamic equilibrium constants of the dissociation of these minerals in aqueous solutions of Na_2S are calculated at 25 to 250°C and 1800bar. The dissolution of HgS results in the formation of HgS_2^{2-} ions whose enthalpy and entropy are calculated. This sulphide complex is significant in the transfer of Hg in hydrothermal solutions under alkaline conditions (Khodakovskii et al, 1975).

Sulphides of Hg (cinnabar, metacinnabarite), Sb (stibnite), As (realgar, orpiment), and pyrite from the active thermal springs of the Kamchatka and Kuril islands have been described in detail by Ozerova et al (1969). Wide variations in Hg content in the samples suggest complex histories of formation.

Around cinnabar deposits, the geochemical zoning found as lateral (or transverse) haloes are in the order: Co, Ag, Ni, Zn, Pb, Cu, Ba, As, Hg. This order indicates that Hg has the widest, and Co the narrowest, halo centred on the deposit.

Some (Beus & Grigorian, 1977) have observed that, in vertical or axial zoning, Hg, together with Ba, Sb, and As, occur near to the surface, followed successively downwards to W thus: Ba - (Hg, Sb, As) - Cd - Ag - Pb - Zn - Au - $Cu_T Bi_T$ - Bi - Co - Mo - U - Sn - Be - W.

MERCURY IN METAMORPHIC ROCKS

The Hg contents in different metamorphic rocks vary widely and range from about 50 to 400ppb, and 158 sample values have been given in table 7 of USGS professional paper, 713 (1970).

The concentration of Hg in the regionally metamorphosed pelitic schists, gabbro rocks and in quartz, magnetite and hematite from metamorphosed iron formation rocks, reported by Jovanovic and Reed (1968) ranged from tenths of one ppm to tens of ppm.

By stepwise heating experiments, they found that the fraction of the total Hg released at higher temperatures, i.e., above 450°C, generally tended to increase with the reported increase in metamorphic grade. An analogous trend has been noted in chondritic meteorites (chapter I). Mercury may thus reflect the thermal history of the sample and serve as a possible geothermometer. The study was made by them with regard to this aspect; exploring the possibility that the observed correlation of $Hg_{<450C}:Hg_{>450C}$ (essentially equal to $Hg_{total}:Hg_{>450C}$) in meteorites is related to the equilibrium conditions established at a given temperature. For this, the regionally metamorphosed pelitic schists were chosen and their thermal history was recorded in mineral assemblages corresponding to the metamorphic grade. Their experimental results showed that, as the degree of metamorphism increased, the relative amount of Hg released at low temperatures decreased while the amount of Hg released at high temperatures increased.

It can be interpreted as arising from an equilibrium process that is regulated by energy differences between the possible high and low temperature sites. Mercury does not seem to have been redistributed as these rocks cooled.

The Hg contents in phyllites, greenschists, staurolite schists and sillimanite gneisses show that progressive dynamothermal metamorphism does not markedly affect the behaviour of Hg (Golovnya & Zlenko, 1970). The Hg contents in these rocks range from 1.5 to 1.9ppm. The migration of Hg in alkali metasomatites occurs in the gas phase. The maximum content of Hg is reported for the sillimanite crystalline schists (7.5ppm). Considerable impoverishment of Hg in high-grade metamorphism is noticed. Ferruginous ore deposits possess a low Hg content. Alkali metasomatites contain redissolved Hg as sulphides. The Hg content in the Pre-Cambrian metamorphosed rocks of the Kursk Magnetic Anomaly and Krivoi Rog ranges from 2.0 to 3.4ppm (Aidinyan et al, 1970).

EFFECTS OF METAMORPHISM

By means of laboratory experiments, it has been shown that the release temperatures of Hg held in the crystal lattice of pyrite and sphalerite are 450 and 600 to 650°C respectively (Watling, 1972). These temperatures correspond to regional metamorphism of the amphibolite and eclogite facies grade, but due to the long periods of time that high temperatures are maintained during metamorphism, mercury present as sulfide grains is released at about 325°C and that adsorbed on crystal surfaces is probably lost below 200°C. A decrease in mercury level was found in a series of rock samples of increasing metamorphic grade from various locations in Ontario.

In sealed-tube experiments on Cu-Hg-S, an increase in the solution of HgS ($Cu_{1.73}S$ - Cu_2S) was found with increasing temperatures between 100 and 400°C (Watling, 1972). These results are regarded as valid for other metal-Hg-S systems, with the further possible implication that Hg can be redistributed in rocks during metamorphism from a silicate to a sulphide phase.

Table 2.2: Mercury in Various Sediments

Locality	Mercury (μg/g)	Reference
Clyde, UK	0.4 - 4.4	Bartlett et al (1978)
Dark, UK	0.01 - 0.55	Dept. Environ. (1976)
Dee, UK	0.02 - 1.3	Dept. Environ. (1976)
Mersey, UK	0.40 - 6.0	Dept. Environ. (1976)
Swansea Bay, UK	0.02 - 1.6	Dept. Environ. (1976)
Plym, UK	0.02 - 2.6	Millward & Herbert (1981)
Rhine, Netherlands	18.0 (average)	De Groot et al (1971)
Ems, Germany	3.0 (average)	De Groot et al (1971)
Loire, France	13.2 - 46.8	Frenet-Robin & Ottmann (1978)
Puget Sound, USA	0.015 - 10.7	Crecelius et al (1975)
Le Have, Canada	0.32 - 3.54	Cranston (1976)

The Outokumpu and Vihanti deposits of Finland reveal the effect of metamorphism on sulphide deposits. The Cu-Zn and Cu-Zn-Pb deposits of Pre-Cambrian rocks have undergone various degrees of regional metamorphism, causing local re-emplacement of ore. The Vihanti deposits of pyrite are enriched proportionally in both Zn and Hg as compared to the undisturbed ore. Remobilised Zn ore exhibits a decrease in Zn grade, but only slightly lower Hg/Zn ratios. At Outokumpu, an increase in the much lower Zn content and almost constant Hg/Zn levels are observed in comparison to the remobilised and original Cu-Zn ore. The data suggest that, if Hg is present mainly in sphalerite, the concentration in this mineral remains constant throughout the remobilisation process. The original Hg content in the Vihanti deposits is still unknown although various data are available. Some ores of Vihanti Zn and pyrite deposits and many other Pre-Cambrian deposits in metamorphic terrains exhibit high concentrations of Hg (greater than 200ppm of mercury).

MERCURY IN SEDIMENTARY ROCKS

The contents of mercury in limestone, sandstone, shale, clay, oceanic and lacustrine sediments, and miscellaneous sedimentary rocks from various locations are presented in tables 8 to 12 of USGS Professional Paper 713 (1970), followed by soil analyses (table 13, ibid).

The mercury contents of various sediments are presented in table 2.2 of this monograph.

The nature of the co-precipitation of mercury with other dissolved elements in the load of mercury deposited in estuary sediments is discussed in chapter IV.

MERCURY AS GEOCHEMICAL INDICATOR

Because of its high mobility and dispersion, mercury is used as a geochemical indicator in exploring for base metal and other deposits. Some noted field observations are given below.

In the USSR, Furzov (1958) collected and analysed nearly 500 samples from the Achsai Pb-Zn deposits. Most of the samples contained less mercury than could be detected by the test employed (0.3ppm). However, mercury haloes were detected in the bed rock of barren limestones overlying a blind Pb-Zn ore deposit. The peak anomaly values varied from 1.2ppm to 6.5ppm. Mercury anomalies were also detected where no Pb or Zn haloes were found, suggesting the gaseous transportation of mercury through the host rock. The mercury anomaly detected in the soils overlying the blind ore was much lower.

Sergeyev (1957) detected haloes, having peaks of 30ppm, over a polymetallic ore deposit. It has been suggested by some authors (Ginzburg, 1960; Furzov, 1958) that the mercury content of a mineral is a function of the temperature of deposition (more explicitly so in the case of sphalerite).

The early work of Grip (1948) and Furzov (1958) showed that mercury is more prevalent in hydrothermal ore deposits than in un-mineralised rocks. In most cases, the mercury content of sulphide ore is of the order of n ppm to 100n ppm, whereas most rocks contain only 0.5ppm or less.

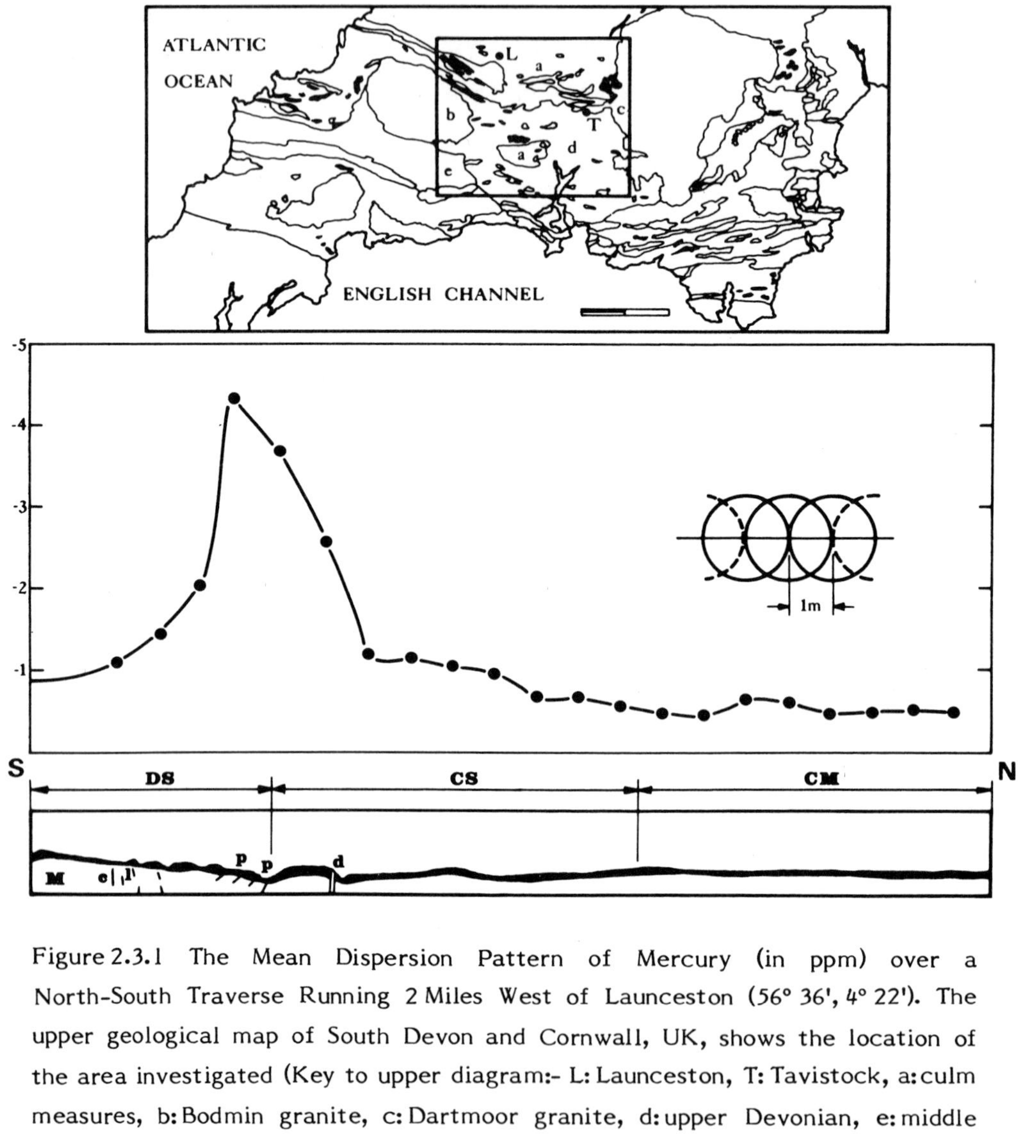

Figure 2.3.1 The Mean Dispersion Pattern of Mercury (in ppm) over a North-South Traverse Running 2 Miles West of Launceston (56° 36', 4° 22'). The upper geological map of South Devon and Cornwall, UK, shows the location of the area investigated (Key to upper diagram:- L: Launceston, T: Tavistock, a: culm measures, b: Bodmin granite, c: Dartmoor granite, d: upper Devonian, e: middle Devonian. The marker is 10 miles long). The average mercury values of stream sediment samples over an overlapping area with a radius of 1 mile were plotted at the centre of the area, as indicated by the inset. (Key to lower diagrams:- DS: upper Devonian slate, CS: carboniferous shale, CM: culm measures, M: mineralised zone, c: Cu, l: Pb, p: pillow lava, d: diabase dyke)

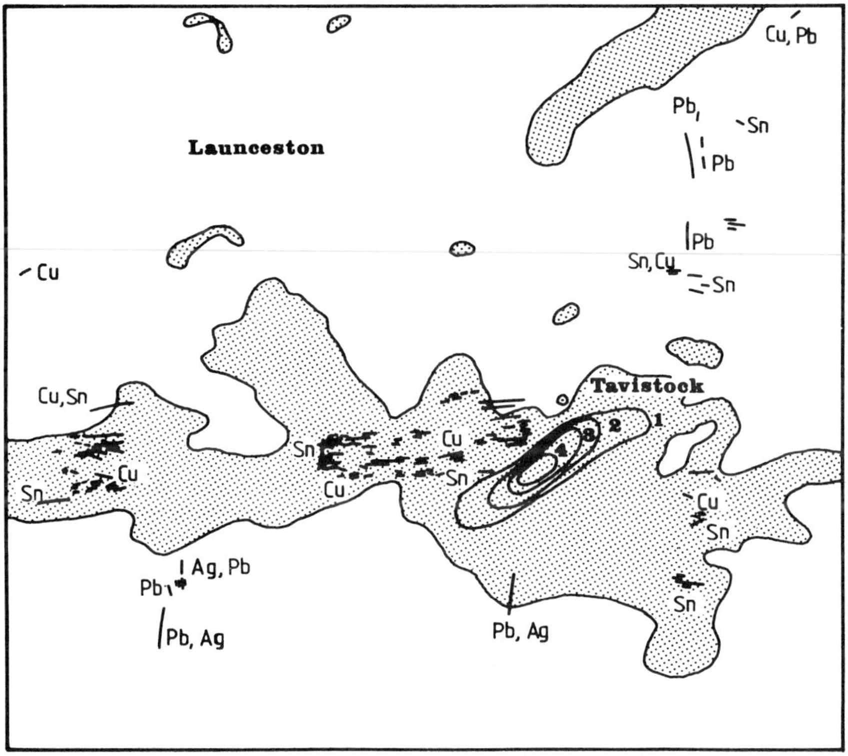

Figure 2.3.2 Residual Mercury Map - Degree One (positive areas shaded)

In several Swedish ore deposits, the concentration of mercury in the ore appears to be related to the Zn content. The results for many ore deposits, including Boliden (Grip, 1948) of North Sweden, show that, if they are divided into Zn-bearing and non Zn-bearing deposits, the ores of the former were found to be higher in mercury than those of the latter.

Ozerova (1959) studied the primary dispersion haloes of mercury on the Pb-Zn deposits of the Turkestan mountain range. The primary haloes over the deposits exhibit values which are ten times as high as those from the barren zones. The dispersion is also ten times higher than the actually mineralised area.

During the sixties, many workers at various mineral deposits identified characteristic mercury haloes (Hawkes & Williston, 1962; James & Webb, 1964; Friedrich & Hawkes, 1966; Warren et al, 1966; Mitra & Webb, 1967; Mitra, 1970; and many others).

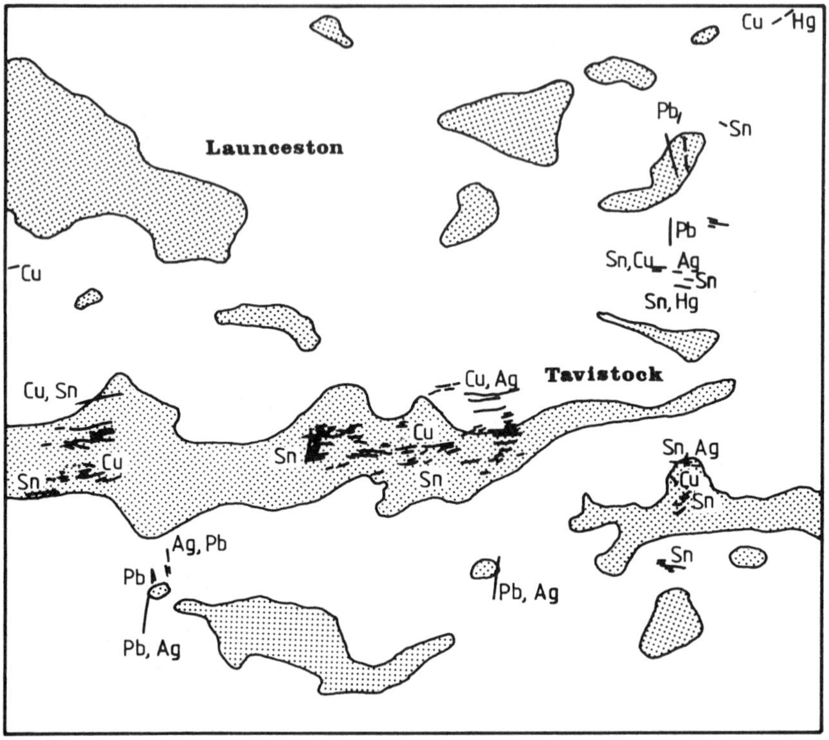

Figure 2.3.3 Residual Mercury Map - Degree Six (positive areas shaded)

Fleischer (1970) reviewed the available Russian literature which presents some anomalously high occurrences of mercury in some rocks and soil samples of several regions of the USSR.

Secondary Dispersion Study for Mineral Exploration: An Example: In a mercurometric survey, stream sediment samples collected from an area of 300 square miles at a density of one sample per square mile around the base metal (Cu, Sn, Pb, Ag) mineralised region near to Tavistock in south-west England, were analysed for mercury (Mitra & Webb, 1967; Mitra, 1970). The regional dispersion of mercury by primary and secondary processes was investigated by using the circular averaging technique (involving overlapping circular sample areas of about 3.14 square miles) developed by Mitra (1970, 1973). A smoothed profile was obtained along a north-south traverse

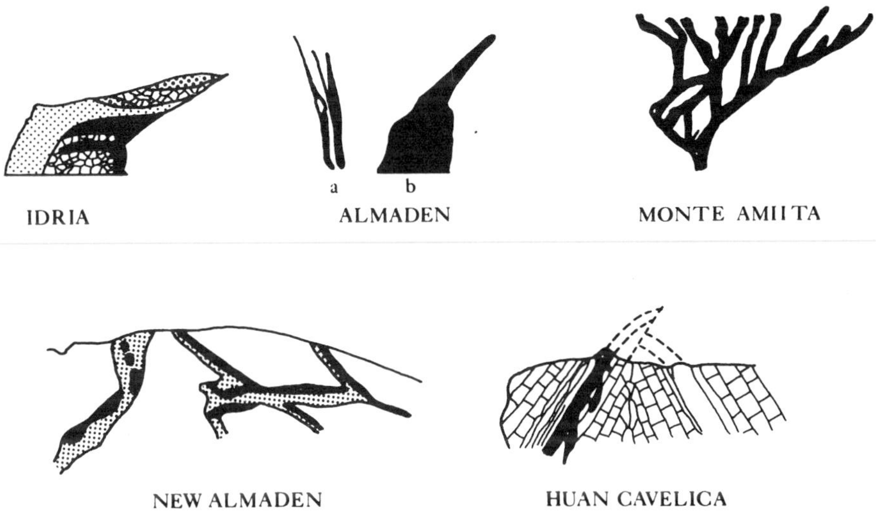

Figure 2.4 Cross-Sectional Views of Some Cinnabar Deposits - Idria (Yugoslavia), Almaden (Spain, a: cross-section, b: longitudinal section), Monte Amiita (Spirek, Italy), New Almaden (California), Huancavelica (Peru)

(running about 3.3km west of Launceston (56° 36', 4° 22') (figure 2.3.1). This technique suppresses the background noise and considerably emphasizes anomalous zones. The dispersion pattern was also studied by the preparation of a series of trend surface and residual maps by computer. The positive residual areas for polynomial surface-fittings of degree one (figure 2.3.2) to degree six (figure 2.3.3) revealed a significant mercury dispersion pattern around the mineralised zones.

MERCURY ORE DEPOSITS

Deposits of mercury occur mainly in volcanically active zones such as the Circum-Pacific volcanic belt, the East Pacific Rise, and the Mid-Atlantic Ridge. On continental land, their distribution is as shown in figure 2.1. Sectional views of some major cinnabar deposits are presented in figure 2.4.

The continental distribution of mercury deposits is discussed below:

EUROPE

In Europe, the main producers are the USSR, Spain, Italy, and Yugoslavia, and the deposits are briefly described below:

Spain is a major producer of mercury. On average, 90% of the mercury production in Spain comes from the Almaden mines. Between 1954 and 1963, the latter produced 20 to 25% of the total world production. The depth of mining has exceeded 3600ft at Almaden and the mine has so far produced about 8 million flasks. The richest ores there contain between 6 and 20wt% of mercury.

The most important deposits are located in an intensely fractured and metamorphosed Silurian Devonian complex of sub-vertical argillites and quartzites. The deposits occur along a major fault line which is known to be mineralised for over 15 miles. The Valley of Rio Valdeazogue runs along a fracture zone which exhibits numerous occurrences of cinnabar at 5.8 and 18km east of Almaden. The deposits lie mainly in a valley.

Three veins constitute the major deposits:

1) San Nicolas, average width of about 10ft
2) San Francisco, average width of about 8ft
3) San Diego (eastern half) and San Pedro (western half) 17 to 36ft wide with ore marking about 60% of this width.

At Mieres (Asturias), mercury ore contains much As.

Italy's position is second in terms of tonnage output, with the Monte Amiita mine and Siels being its major sources. The ores of Monte Amiita have a mercury content of between 0.6 and 3%. The deposits are located around an extinct volcano. They are found concentrated along the fractured contact zone running between Mesozoic and Cenozoic limestones and shales. Antimony is often found in an associated state.

In Yugoslavia, the Idria mine near to Trieste has been operating since the fifteenth century and has supplied an amount which is equal to about one third of the total output of the Almaden mines. At Idria, the cinnabar deposits occur above a thrust plane in the Carboniferous and Triassic sandstones, limestones, and schists. Mineralisation is found mainly in the fractured zones.

The USSR is another major producer. At Nikitovka (Donets Basin) in the Ukraine, cinnabar occurs in a carboniferous sandstone bed, dipping at 50° below shales. Other deposits are recorded in the Caucasus and in the Fergana in silicified limestone.

Features of the Altai deposits include a strong tectonic control and consanguinity with Pb, Zn, and Cu mineralisation. Mercury also occurs in the Maritime Provinces. About 200 deposits are known, and the USSR is an exporter of mercury.

In Turkey, the mercury deposits are concentrated in the Izmir and Kutahaya provinces.

USA

In the USA, Humboldt county in Nevada, together with California, is the major domestic producer, with Texas in second place. The estimated reserves of the US amount to 1.5 million flasks. A 30 year supply is foreseen at the present consumption rate of 60000 flasks per year and selling price of US$500 per flask.

In New Almaden (California), cinnabar deposits occur as stockworks in silicified and fractured serpentines and metamorphic rocks along two deep fractures. The other major mine in California is located at New Idria, where the ores lie along clay patches between serpentine and sandstone. The ores occur as veins, irregular stockworks, and impregnations. Meta-cinnabar is a common associate.

In Texas, the Terlingua deposits occur in three zones:

a) at limestone-clay contacts, b) in Karstic cavities, c) in breccia pipes, and d) in igneous rocks ranging from rhyolite to syenogabbro.

In Alaska, the Red Devil mine is located on the western limits of an upper Cretaceous anticline, which has been subjected to several stages of faulting. Two perpendicular andesite dykes are the foci of mineralisation. About 25 ore shoots are located in ancient openings, due to small faults. The ore is complex: the stibnite - realgar - orpiment content exceeds the cinnabar content.

SOUTH AMERICA

In Mexico, the Huitzuco deposit lies in a strongly fractured limestone crossed by intrusive granite. The minerals are livingstonite, cinnabar, meta-cinnabar, and antimonite. The Cuadalczar deposit is another important source of mercury in Mexico.

Chile and Columbia

Here, production comes from "Los Mantos" (near to Punataqui, Chile) and "La Experanza" (Colombia).

In Peru, the Hauncavelica deposit is located along a north-south trending anticline cut by tertiary andesites and basalts. The cavities and pores are filled with cinnabar, realgar, etc.

ASIA

In China, the province of Kveichow, with the surrounding provinces of Szechuan, Hunan, and Kwangei, is a long-time production centre for mercury. Fractured anticlines are the preferred foci of ore deposition. The host rocks are Paleozoic limestones and shales.

In Japan, the important mines are located at Itomuka and Yamato.

AUSTRALASIA

In the Phillipines, the most important deposits are located on the Lalawan Islands.

WORLD PRODUCTION AND MAJOR USES

Since 1850, most of the world's mercury has come from four districts: Almaden (Spain), Idria (Yugoslavia), Monte Amiita (Italy), and California (USA). Before Italy and the USA came to prominence, Santa Barbara in Peru was one of the major producers of mercury, together with Almaden and Idria.

The world production (number of 76lb flasks) of mercury by country is shown in table 2.3. Since the seventies, the USSR has been at the top of the production league. The top ten producers account for over 90% of the world's mercury supply. The principal exporters of mercury during 1975-1976 were Spain, Italy, Yugoslavia, Mexico, Algeria, the USSR, and the Peoples' Republic of China.

Upsurges in mercury production occurred during the Korean War and again during the Vietnam war. The present decline in production is due to excessive stockpiling by the US, to a sluggish economy, to environmental control measures, and to an increasing degree of recovery from secondary sources (e.g. electrical apparatus, spent mercury batteries, industrial and control instruments, mercury boilers, etc). The secondary production accounts for between 20 and 30% of the total consumption in the USA. Only 20% of the mercury used annually in the OECD countries is recycled (OECD report, 1974).

The USA is the largest consumer, with imports from Italy, Algeria, Yugoslavia, Spain,

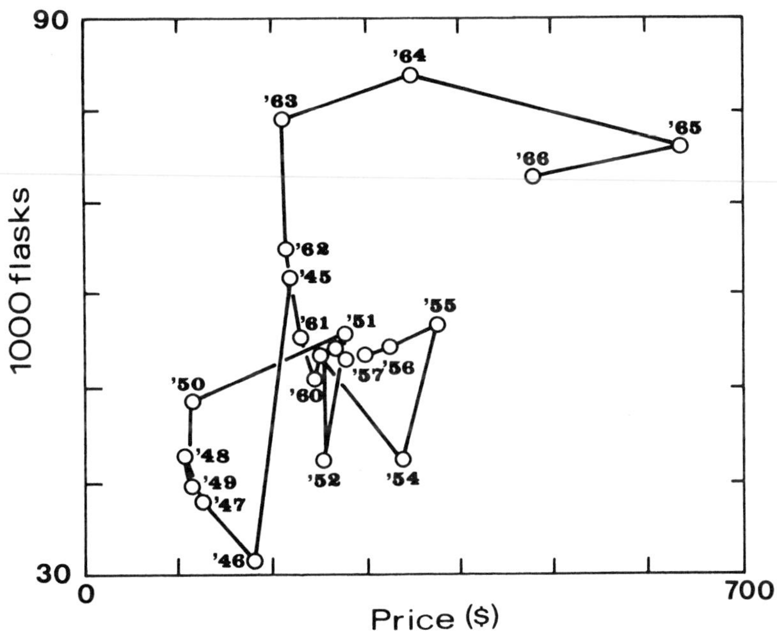

Figure 2.5 Consumption and Price Trends, Illustrating Absence of Correlation. (Price at New York, expressed in terms of the 1968 value of the dollar)

the Peoples' Republic of China, Canada, and Mexico. The USSR is the next largest consumer but the demand is entirely satisfied by her own domestic supply.

PRICE

The price of mercury has fluctuated erratically since the Second World War. A depression of the price during 1947-49 was due to the release of an accumulated stock of mercury by Germany and Japan which came onto the market after the war was over. The price increased and peaked in 1955, during the Korean war, at US$290 per flask. The price decreased until 1963 and then rose again to reach a height of about US$750 per flask in mid-1965 during a bad phase of the Vietnam war.

The consumption of mercury is not always related to its price. This lack of correlation between the price and production in the US is illustrated by figure 2.5.

Table 2.3: World Production of Mercury (thousands of 76lb flasks) by Country
[Data from George (1977), Minerals Yearbook]

Country	1925	1930	1935	1940	1945	1950	1955	1960	1965	1970	1975
Africa											
Algeria	0.055	0.325	-	0.791	0.326	-	-	-	-	-	13.3
South Africa	-	-	-	0.042	0.852	-	-	-	-	-	-
Asia											
China	0.095	0.725	1.313	3.403	1.828	1.450	11.500	23	26	20	26
Japan	-	0.121	0.148	3.394	3.139	1.312	4.99	5.791	4.536	5.17	-
Australasia											
Phillipines	-	-	-	-	-	-	0.635	3.041	2.384	4.648	0.232
Europe											
Czechoslovakia	2.129	2.06	2.004	2.584	0.435	0.725	0.725	0.725	0.73	4.815	6
Italy	53.189	56.069	28.191	91.230	25.410	53.346	53.52	55.492	57.32	44.47	30.4
Spain	37.052	19.221	35.559	52.214	40.694	51.808	36.231	53.369	82.76	45.552	47.051
Turkey	0.098	0.537	0.025	0.5	0.158	-	0.841	1.339	2.755	8.592	8.8
USSR	0.287	3.278	8.7	8.7	7.3	11.6	12.3	25	40	48	55
West Germany	-	-	0.116	0.957	-	-	-	-	-	-	7
Yugoslavia	-	-	-	-	8.876	14.649	14.516	14.069	16.419	15.461	16.941
North America	-	-	-	-	-	-	-	-	0.02	24.4	12
Canada	-	-	2.024	-	-	-	-	-	-	-	-
USA	9.053	21.553	17.518	37.777	30.763	4.535	18.955	3.223	19.582	2.7296	7.366
South America	-	-	-	-	0.003	-	-	-	0.052	0.012	-
Bolivia	-	-	0.25	-	-	-	-	-	-	-	-
Chile	-	-	-	0.1	0.862	0.314	0.526	2.876	0.428	0.388	-
Colombia	-	-	-	-	-	-	-	-	0.046	0.215	-
Mexico	1.123	4.946	6.277	11.653	16.443	3.757	29.881	20.114	19.190	30.256	14.184
Peru	-	-	-	-	0.209	-	0.148	3.034	3.117	3.196	3.500
Other countries	-	-	-	-	-	-	-	31	0.66	1.5	0.652
World Total	103	109	100	216	137	143	185	242	276	284	249

USES

The unusual physico-chemical properties of Hg have made it useful for many purposes (table 2.4).

Cinnabar (HgS) has been shown to have been smeared on neolithic skulls found in Italy. In about 2000BC, mercury pigment was used on a tomb on the island of Syria in the Mediterranean (Goldwater, 1972). However, the uses of mercury have changed drastically during the last four centuries. Nowadays, the number of applications exceeds three thousand in number (Bailey & Smith, 1964).

Table 2.4: Physico-Chemical Characteristics and Major Uses of Mercury

Characteristic	Exploited in
Fluidity at ordinary temperatures High surface tension Uniform volume expansion No wetting of glass	Barometers Manometers Thermometers Various measuring devices
High electrical conductivity	Electrical conductors, switches
High thermal conductivity	Coolant
High thermal neutron capture cross-section (360 barn)	Neutron shield in atomic devices
Ability to form amalgams with many metals	Dental fillings and metal recovery processes
Strongly toxic compounds	Bactericides, fungicides, insecticides, and pharmaceuticals
Brilliantly hued compounds	Paints

The early part of this century saw Hg being used mainly in the recovery of gold and silver, and in the making of fulminate and vermillion. Today, some 50% of mercury usage is in electrical apparatus and in the production of caustic soda and chlorine.

REFERENCES

- Adinyan, N.K., Ozerova, N.A., and Glovnya, S.V. (1970) - Ocherki, Geokhimiya Rtuti, Molybdena, Sery, Gidrotermichesky Protsess, 1970, 70-80
- Baev, V.G. (1968) - Doklady Akademii Nauk SSSR, 181[5], 1249-51
- Baev, V.G. (1974) - Tezisy, Doklady Konferentsii Geologii Poleznuikh Iskopaemuikh Severnogo Kavkaza, 4th, 136-7
- Bailey, D.S. (1975) - in International Conference on Heavy Metals in the Environment, Toronto, C25-31
- Bailey, E.H. and Smith, R.M. (1964) - Mercury, Its Occurrence and Economic Trends, US Geological Survey Circular 496, 8-9
- Bartlett, P.D., Craig, P.J., and Morton, S.F. (1978) - Science of the Total Environment, 10, 245-51
- Beus, A.A. and Grigorian, S.V. (1977) - Geochemical Exploration Methods for Mineral Deposits, Applied Publishing Co., Wilmetter, Illinois, 287
- Burykhin, I.V. and Burykhina, Z.E. (1974) - Geologiya Poleznuikh Iskopaemuikh Kirgiza, 87-94
- Carr, R.H., Jones, M.M., Warner, T.B., Cheek, C.H., and Russ, E.R. (1975) - Nature, 258, 588-9
- Cranston, R.E. (1976) - Estuarine and Coastal Marine Science, 4, 695-700
- Crecelius, E.A., Bothner, M.H., and Carpenter, R. (1975) - Environmental Science and Technology, 9, 325-33
- Degens, E.T. and Ross, D.A. (1969) - Hot Brines and Recent Heavy Metal Deposits in the Red Sea, Springer-Verlag, New York, 600
- De Groot, A.J., De Geoij, J.J., and Zegers, C. (1971) - Geologie Mijnbouw, 50, 393-8
- Department of the Environment (1976) - Environmental Mercury and Man, Pollution Paper, No.10, London, HMSO p92

- Dissanayake, C.B. and Vincent, E.A. (1975) - Mineralogical Magazine, 40, 33-42
- Donovan, P.R. and James, C.H. (1967) - Geological Survey of Canada, Paper 66-54, 89-110
- Eshleman, A., Siegel, S.M., and Siegel, B.Z. (1971) - Nature, 233, 471
- Ferm, R. and Larsson, J.E. (1973) - Kricksilver anvandring, Kontroll och Miljoeffekter Statens Naturvardsvesk, PM421, 8-17
- Flanagan, F.J., Moore, R., and Aruscavage, P.J. (1982) - Geostandards Newsletter, 6(1), 25-46
- Frenet-Robin, M. and Ottmann, F. (1978) - Estuarine and Coastal Marine Science, 7, 425-36
- Friedrich, G.H. and Hawkes, H.E. (1966) - Economic Geology, 61[4], 744-53
- Furzov, V.Z. (1958) - Geokhimiya, 3, 388
- George, J.G. (1970-77) - Mercury. Canadian Minerals Yearbooks, Publishing Centre, Department of Supply and Services, Ottawa, Ontario
- Ginzburg, I.I. (1960) - Principles of Geochemical Prospecting (trans), Pergamon Press, London
- Goldschmidt, V.M. (1937) - Journal of the Chemical Society, 655-73
- Goldschmidt, V.M. (1954) - Geochemistry, Oxford University Press, Baltimore, Maryland
- Goldwater, L.J. (1972) - Mercury, a History of Quicksilver, York Press, Baltimore, Maryland
- Golovnya, S.V. and Zlenko, B.P. (1970) - Ocherki Geokhimiya Rtuti, Molybdena Sery, Gidrotermichesky Protsess, 1970, 81-8
- Grigorian, S.V. (1974) - Sovetskaya Geologiya, 1973[1], 15-33 (English translation, 1974)
- Grip, E. (1948) - Sveriges Geologiska Undersokning Arsbok, 42C[8], 499
- Hawkes, H.E. and Williston, S.H. (1962) - Mining Congress Journal, 48[12], 30-2
- Heyl, A.V., Broughton, W.A., and West, W.S. (1970) - Natural History Survey Internal Circular, 19, 49
- Heyl, A.V. (1972) - Economic Geology, 67, 879-94
- James, C.H. and Webb, J.S. (1964) - Transactions of the Institute of Mining and Metallurgy, 73pt, 9, 1963-64

- Jonasson, I.R. and Boyle, R.W. (1972) - in Proceedings Soc. Canada Symposium on Mercury in Man's Environment, Ottawa, 5-21

- Jovanovic, S. and Reed, G.W. (1968) - Geochimica et Cosmochimica Acta, 32, 341-6

- Khodakovskii, I.L., Popova, M.J., and Ozerova, N.A. (1975) - Geokhimiya, 1975[3], 360-70)

- Kuznetsov, V.A. (1972) - Geologiya Genexis Endogennykh Rud Sibir, 23-52 (Russian)

- Kuznetsov, V.A. (1974) - Geologiya i Geofizika, 1974[5], 103-22

- Kuznetsov, V.A. and Obolenskii, A.A. (1969) - Rudoobrazov. Ego Suyaz. Magmat. Soveshch (Doklady) (published 1972), 137-44 (Russian)

- Kuznetsov, V.A., Plekhova, K.R., and Lyapicheva, A.A. (1974) - Soviet Geology, 1974[2], 142-6 (Russian)

- Menyaylova, G.A. (1967) - Doklady Akademii of Sciences, USSR, Earth Science Section, 174, 150-2

- Millward, G.E. and Herbert, I. (1981) - Environmental Pollution (Series B), 2, 265-74

- Minerals Yearbook (1950-77) - Bureau of Mines, US Government Printing Office, Washington

- Mitchell, A.H.G. and Bell, J.D. (1973) - Journal of Geology, 81, 381-405

- Mitchell, A.H.G. and Garson, M.S. (1976) - Mineral Science Engineering, 8[2], 129-69

- Mitra, S. (1970) - Proceedings of the Indian National Science Academy, 36A[4], 191-210

- Mitra, S. (1973) - in Indo-Soviet Symposium on Recent Trends in the Exploration of Minerals, Oil, and Groundwater, New Delhi, October, 1973 (abstract)

- Mitra, S. and Webb, J.S. (1967) - Transactions of the Institute of Mining and Metallurgy, 76B, 180

- Mogarovskii, V.V., Melnichenko, A.K., and Klimov, G.K. (1975) - Doklady Akademii Nauk Tadzhiskoi SSR, 18[6], 44-7

- Novokhatsku, M.P. and Kalinin, S.K. (1952) - Vestnik Academii Nauk Kazakhskoi SSR, No.9

- OECD Report (1974) - Mercury and the Environment, studies of Mercury Use, Emission, Biological Impact and Control, Paris, 147-70)

- Ozerova, N.A. (1959) - Geokhimiya, 6, 38-45

- Ozerova, N.A. (1962) - Proc. Geol. Ore Deposits, Petrog. Min. Geochem. (Chikrov Inst. IGEM), No.72, Questions of Geochemistry, Pt 4, Academy of Sciences of the USSR. Akademy of Science Press, Moscow, USSR

- Ozerova, N.A. and Aidinyan, N.K. (1970) - in Problems of Hydrothermal Ore Deposition, Inst. Union Geol. Sci. A, No.2, Stuttgart

- Ozerova, N.A., Aidinyan, N.K., and Dobrovolskaya, M.G. (1969) - Vulkanizm Glubiny Zembli. Mater, Vses.

- Preuss, E. (1941) - Zeitschrift Angewandte Mineralogie, 3, 8

- Rankama, K. and Sahama, T. (1950) - Geochemistry, Chicago University Press, Chicago

- Saukov, A.A. (1946) - Geokhimia Rtuti, Trudu Instituta Geologicaeskogo, Academii Nauk SSSR, 78[17], 129

- Saukov, A.A., Aidinyan, N.K., and Ozerova, N.A. (1972) - ed. "Nauka", Moscow, (Russian)

- Sears, W.P. (1971) - Proceedings of the Third International Geochemical Exploration Symposium, Canadian Institute of Mining and Metallurgy (Edited by R.W.Boyle), Special volume No. 21, 384-90

- Sergeyev, J.A. (1957) - Proceedings of the First All Union Conference on Geochemical Prospecting for Ore Deposits (V.I.Krasnikov), University of Alabama

- Shashkina, V.P., Oleinik, D.V., and Nikolskaya, I.N. (1972) - Voprosy Geologii Rudnuikh Mestorozhdenuikh Dalnego Vostoka, 141-54 (Russian)

- Sidgwick, N.V. (1950) - The Chemical Elements and their Compounds, Vol.2, Oxford University Press, London

- Smith, E.C. and Webber, G.R. (1972) - Proceedings of the International Symposium on Geochemical Exploration in Canada, 71-80

- Stakeev, J.I., Gerasimovsku, V.I., and Stakheeva, S.A. (1975) - Geokhimiya (II), 1668-75

- Stock, A. and Cucuel, F. (1934) - Naturwissenschaften, 22, 390-3

- Taylor, S.R. (1964) - Geochimica et Cosmochimica Acta, 28, 1273-86

- US Geological Survey (1970) - Professional Paper 713, 67pp

- Ushak, A.T. and Trufanov, V.N. (1968) - Eksp. Issled. Mineraloobrazov, Mater. Vseb. Voveshch. Eksp. Tekh. Petrogr. 8th, 1968 (published 1971), 52-5, (Russian)

- Uzgalis, E. (1975) - Soviet Geology (12), 80-90
- Warren, H.V., Delavault, R.E., and Borakso, J. (1966) - Economic Geology, 61[6], 1010-28
- Watling, R.J. (1972) - Transactions of the Institute of Mining and Metallurgy (Section B), 81, 47-8
- Watling, R.J., Davis, G.R., and Meyer, W.T. (1973) - London Symposium, Vol. 59-70
- Webber, G.R. and Newbury, M.L. (1971) - Canadian Journal of Earth Sciences, 8, 1197-202
- Wedepohl, K.H. (1969-78) - Handbook of Geochemistry, Volumes 2-4, Springer-Verlag, Berlin
- White, D.E. (1967) - in Geochemistry of Hydrothermal Ore Deposits, (Edited by H.L.Barnes), Hold Reinhardt & Winston, New York, 479-528
- White, D.E. (1974) - Economic Geology (Bulletin of the Society of Economic Geologists), 69[6], 954-73

CHAPTER III

SOURCES OF MERCURY IN THE ENVIRONMENT

INTRODUCTION

Many models have been used to calculate the amounts and movement of mercury in the world (Weiss et al, 1971; Kothny, 1973; Miller & Buchanan, 1979).

A large proportion of the mercury found in the environment is derived from industrially produced mercury and amounts to approximately 10000 tons per year. In the northern hemisphere, higher concentrations of mercury have been observed in air (Slemr et al, 1981), in sea water (Gardner, 1975), and in the hair of the people. The industries which are mainly responsible for the dispersion of mercury are the burning of fossil fuels, compost incinerators, mining and extraction of mercury from cinnabar, the chlor-alkali industries, paper pulp, paints, fungicides, electrical equipment, instrumentation, amalgamation, etc. Mercury droplets from broken thermometers can also be a source of exposure to mercury vapour in the home and in hospital.

Other sources of mercury are the burning of fossil fuels and ores. Although the concentration of mercury in fuels is small, the latter are consumed at an enormous rate and therefore constitute a significant source of mercury in the environment. The release of mercury from coal used all over the world amounts to between 1.4×10^8 and 2.72×10^9 g/year, depending upon the mercury concentration in the coal and the amount of mercury trapped by precipitators before release from chimney stacks.

The mercury content of ores (excluding mercury ores proper) is likely to be higher than that of coals. However, the tonnage of ores mined annually is much lower than that of coal. The upper limit for the natural release of mercury due to chemical weathering can be estimated by comparison with the corresponding figures for sodium, and it has been assumed that about 230 tons of mercury per year are leached out (Joensuu, 1971).

Table 3.1: Uses and Consumption of Mercury in the United States (thousands of 76lb flasks. Data from Minerals Yearbook, USGS)

Use	'42	'45	'50	'52	'55	'60	'65	'70	'71	'75
Agriculture (a)	1.5	2.9	4.5	7.7	7.7	3.0	3.1	1.8	1.5	0.6
Amalgamation (b)	0.2	0.2	0.2	0.2	0.2	0.3	0.5	0.2	0	0
Catalysis	-	3.7	2.7	0	0.7	1.0	0.9	2.2	-	0.8
Chlor-alkali (c)	0.5	0.6	1.3	1.5	3.1	6.2	8.8	15.0	12.3	15.2
Dentistry	2.1	0.5	1.5	1.6	1.4	1.8	1.6	2.3	2.4	2.3
Electrical (d)	5.5	24.5	12.0	13.2	6.5	9.6	14.8	16.0	16.9	17.0
Felt industry (e)	0.2	0	0	0	0	0	0	0	0	0
Instruments	6.6	3.3	5.4	10.4	5.4	6.5	4.6	4.8	4.0	4.6
Laboratories	0.3	0.3	0.6	0.5	1.0	1.3	2.8	1.8	0.18	0.3
Munitions (f)	7.8	1.1	0.1	0.5	0.1	0	0	0	0	0
Antifouling paint (g)	1.2	1.6	3.1	2.5	0.7	1.4	0.3	0.2	0.4	-
Anti-mildew paint	0	0	0	0	0	2.7	7.5	10.1	8.2	6.9
Paper and pulp (h)	0	0	0	0	0	3.5	0.6	0.2	0	0
Pharmaceuticals	8.1	11.1	6.0	2.0	1.6	1.7	3.3	0.7	0.7	0.5
Vermillion (i)	0.2	0	0	0	0	0	0	0	0	0
Redistilled	-	9.6	7.6	-	9.6	9.7	12.3	-	-	-
Total	49.8	62.2	49.2	56.8	54.7	51.5	76.5	61.5	52.5	50.8

a) germicides, fungicides, algaecides, b) major use before 1900, c) industry started in 1937, d) mostly batteries, e) none since 1943, f) blasting caps etc. (none since 1956), g) for ships' bottoms, h) slimicide, i) none since 1944

If the average mercury content of crude oil is taken to be about 0.1ppm, then some 200 tons of Hg can be expected to be released into the atmosphere annually due to the combustion of petroleum alone.

Another common source of mercury hazard is the use of mercury in dental practice. In the USA alone, 100000kg of mercury are believed to be used each year in dentistry for the preparation of dental amalgam restorations (Rupp & Paffenbarger, 1971). The use and consumption of mercury in the USA is shown in table 3.1.

From the known average mercury content of snow or rainwater, it has been estimated that the surface of the earth receives annually some 100000 tons of mercury via precipitation from all of the mercury sources, including natural ores.

The problem of mercury pollution has arisen due only to the uneven distribution of the release of mercury. Some 70 million tons of mercury are present in the sea and, each year, about 20000 tons are precipitated onto the earth's surface by rainfall (Korringa & Hagel, 1973). Therefore, a restriction must be imposed on mankind's activities so that the uneven distribution of mercury is minimised. Only by taking such steps, can one achieve immunity to mercury pollution.

The use of grey powders (chalk and mercury in a ratio of 2 to 1) for latent fingerprint development has been held to be responsible for mercurialism in several detectives (Forbes & White, 1952). In earlier days, the mercury nitrate process for the felting of the fur used in the manufacture of hats led to stomatitis (inflammation of the mouth), salivation, tremors, and muscular atrophy (Neal et al, 1941).

THE MAJOR SOURCES

a. CHLORINE INDUSTRY

Nearly all chlorine is manufactured by the electrolysis of brine, using mercury cells in which the cathode is a flowing sheet of elemental mercury.

$$\text{NaCl (solution)} \xrightarrow{\text{electrolysis}} \text{Cl}_2\text{ (anode)} + \text{NaHg}_x\text{ (cathode)}$$

$$\text{NaHg}_x + \text{H}_2\text{O} \rightarrow \text{NaOH} + x\text{Hg} + (1/2)\text{H}_2$$

A typical, modern $30m^2$ Hg-cell can contain up to 12000lb of Hg which is circulated in a closed system and re-used indefinitely. However, due to the circumstances of its operation, there is a loss of 150-250g of Hg per kg of chlorine produced. "Chlor-alkali Mercury Regulation" (US) aims at reducing this release of mercury. The caustic soda and potash produced by the Hg-cell process contain about 4-5ppm of Hg (Study Group on Mercury Hazards, 1970). Since chlorine, caustic soda, and potash are widely used in other industrial processes and products, the chlor-alkali industries are a potential source of mercury in the environment. Also, all of the hydrogen gas produced during the preparation of chlorine by the Hg-cell process contains mercury. Without proper monitoring, as much as 50lb of Hg may be emitted per 100 tons of chlorine produced.

In Sweden, discharges from the Hg-cell process in 1967 amounted to between 51 and 85g of Hg per ton of chlorine produced. This resulted in a discharge of 30-40g to water, 5-10g to hydrogen gas, 1-10g to the caustics, and 15-25g to ventilation (Study Group on Mercury Hazards, 1970). However the Swedish government succeeded, in 1970, in reducing the total discharge of mercury to about 0.53 to 1.11g per ton of chlorine produced. This was made up of a discharge of 0.01-0.1g to water, 0.01g to hydrogen, 0.01-0.5g to caustics, and 0.5g to ventilation (Study Group on Mercury Hazards, 1970). In Canada, a loss of 0.5lb of Hg per ton of chlorine produced by the Hg-cell process was reported by Fimreite (1970). In the USA, a drastic reduction in the loss of mercury in the chlorine industry has been achieved, as demonstrated by a reduction in the consumption of mercury by the industry from 20770 flasks in 1960 to 14977 flasks in 1970.

Smith et al (1970), in a study of chlor-alkali workers, found that only 17% of them had an average exposure of more than $100 \mu g/m^3$ and suffered from tremors.

b. PAPER AND PULP

Organic mercury compounds (mainly phenylmercury) are often used extensively in the pulp and paper industry. Bouveng (1967) found that some 5 to 20% of the mercury is discharged into waterways, with the rest remaining in the product and finally being released into the environment, e.g., by the burning of paper in garbage. Caustic soda is also used in the pulp and paper industry and, therefore, the caustic soda produced by the Hg-cell process can contribute significant quantities of mercury to paper products. In the USA, the use of organic mercurials as slimicides in the paper industry has been reduced almost to zero since 1973 by legislation.

c. MINING AND REFINING

Upon heating, a large number of minerals and ores release mercury impurities into the air (WHO, 1976). The world release of mercury from ore processing is more than 2000 tonne/year (Weiss et al, 1971). For iron ores, the mean value is 0.09ppm, whereas for calcite it is 0.01ppm, and for barytes it is 0.73ppm (range: 0.02 to 200ppm, D'Itri, 1973).

The discharge of tailings from the mining of cinnabar and other metal sulphide ores can contribute significant amounts of mercury to waterways. When, in the process of refining, sulphide ores are heated in retorts or furnaces, a significant discharge of mercury into the air occurs. In the refining of mercury ores, stack losses of mercury are usually 2-3%. Stahl (1969) has estimated an emission of 50000lb of Hg from mercury smelting, assuming a 3% stack loss. Furthermore, it has been pointed out that the refining of other sulphide ores releases Hg vapour into the atmosphere. Previously, considerable quantities of Hg were used to recover free gold and silver from placer and lode ores in gold and silver mining. Mercury amalgams are used in master mould and investment casting. In jewelry mould casting, the levels of Hg in the air may be as high as 1200 $\mu g/m^3$, and urine levels as high as 2090 $\mu g/l$ in workers have been noted (Copplestone & McArthur, 1967).

The atmosphere at ground level near to mercury ore deposits can contain as much as 20000 $\mu g/m^3$ of Hg. On a weight basis, this is almost 16ppb. High concentrations of Hg are likely to occur in underground water because of the longer and more intimate contact with mineral grains, and other environmental factors. Limited sampling of oilfield brines in California has shown that these contain between 100 and 200ppb of mercury. In hot springs in the same state, the values ranged from 0.5 to 3.0ppb. In one sample, values as high as 20ppb were recorded. Vapours issuing from fumaroles, and steam condensing from hot springs, were reported to have mercury concentrations of 6ppb and 130ppb, respectively. Dust from sulphide-bearing mineral deposits may occasionally be a significant local source of mercury, so much so that dust produced during the treatment of tin ores has been used for the industrial recovery of mercury. Underground cinnabar mining is relatively less hazardous than mining shale for its native mercury content. Exposure is at a maximum during the refining operation.

Unpolluted air contains less than 1 to perhaps 10ng/m^3 of mercury. "Natural pollution" caused by the volatilisation of mercury from ore deposits of mercury or base metals gave values of up to 62ng/m^3. Very large amounts of mercury reach the atmosphere from volcanic emanations, including those from mud volcanoes.

d. PAINT

Organo-mercurial compounds: oleate, phenylmercury compounds and dodecenylsuccinate are used as bactericides and fungicides in the paint industry.

Mercury may be released from the manufacture of paint and also from the use of paints containing mercury by 1) waste water discharges, 2) volatilisation of mercury from painted surfaces, 3) slow release of mercury from anti-fouling paints on ships' bottoms, 4) discharges of mercury into sewers from the washing of paint brushes, rollers, containers, etc. Hanson (1971) has reported an increased mercury level in fish stored in cages in localised areas, due to release from anti-fouling paints into the water. In the vicinity of shipyards undertaking the reconditioning of ships previously painted with Hg-based anti-fouling paints, increased levels of mercury were observed. In the USA in 1976, mildew-proofing alone accounted for 7846 flasks of mercury.

e. LABORATORIES, HOSPITALS, AND DENTAL APPLICATIONS

Mercury is used in commercial laboratories, hospitals, and dental clinics in several forms, including: drugs, reagents, dental preparations, disinfectants, and sterilising solutions. It is used in barometers, manometers, thermometers, porosimeters, coulometers, diffusion pumps, Toepler pumps, air pumps, pump seals, mercury jet electrodes, dropping electrodes, McLeod gauges, Western standard cells, fuel cells, and as a vibration damper, coolant, radiation shield, ballast in emergency flotation and rescue submersibles. These uses undoubtedly result in discharges to municipal sewage treatment plants. It has been estimated by the Bureau of Food and Drug Administration, and departments of Health, Education, and Welfare, that 25% of the approximately 200000lb of Hg per year used in dental preparations, i.e. 50000lb, is lost as particles of amalgam which are scraped off the tooth or fall into the mouth, are spat into cuspidors, and are finally released into sewers (Mercurial Pesticide Registration Review Panel, 1971). Approximately 50% (100000lb) of the total annual use of mercury in dental preparations is actually applied to the teeth. The exposure of patients to mercury was studied by Frykholm (1970), who placed 4 or 5 amalgam fillings in each of 5 patients and used a radio-isotope of mercury in preparing the amalgam. On the fifth day after emplacement of the fillings, the urinary excretion of tagged mercury had gradually increased in each individual and averaged 2.5mg/l. By the seventh or eighth day, the tagged Hg level had dropped to zero in each case. The amalgam fillings containing radio-active mercury were all removed in order to avoid further exposure to radiation when the urinary excretion of tagged mercury reached zero. On the day after removal, the urine level of tagged mercury increased to 5μg/l and then dropped to zero within two days.

By the use of radioactive ^{203}Hg, Frykholm (1970) studied the penetration of mercury through hard dental tissues. These observations, both in vitro and in vivo, on human, monkey, and dog teeth revealed the presence of some tenth of a microgram in the pulp tissue of individual teeth. No tagged mercury was detected in organs remote from the

oral region. Spectrographic analyses of dentin revealed the presence of mercury, silver, tin, copper, and zinc. It was also observed that zinc oxide eugenol and zinc phosphate cement bases blocked the penetration of these metals.
Soremark (1968) used ^{197}Hg in amalgams and observed that it penetrated dental tissues. They also observed the blocking of this penetration by zinc oxide eugenol cement. In addition, they reported that cavity liners prevent this efflux of metal ions into the dentin.
The total mercury level in the air of a dental laboratory can reach $12000 \mu g/m^3$ during the removal of dental amalgam fillings. Iyer et al (1976) noted a sensory neuropathy in a dentist which resolved upon chelation therapy.
An area of concern is contamination of the environment by the disposal of waste amalgam. Each time that a dentist uses an amalgam filling, he must prepare an excess in order to ensure proper execution of the restorative work. The excess amalgam removed during shaping of the restoration is usually rinsed from the mouth or aspirated, and it should be caught in a strainer or trap in the waste drain. Unfortunately, appropriate solid waste traps are not present in many cuspidors, aspirators, and evacuators. A survey of several geographical areas in the US revealed that recovery of amalgam scrap is widespread. In 1969, the Department of Defense recovered, from the Army, Navy, and Air Force dental corps, 42000lb of scrap amalgam. In hospital laboratories, Hg levels as high as $710 \mu g/m^3$ have been noted.

f. SEWAGE TREATMENT PLANTS

Raw sewage contains an average of 2ppb of Hg, of which 1ppb comes from the effluent of dental clinics, laboratories, and hospitals (US EPA, 1975).
Mercury may be released to the waste water treatment plants by a) mechanical shredders (comminutors) using mercury seals, b) flow meters and flow-rate controllers, and c) rotary sewage flow distributors using mercury seals. An estimate by the Environmental Protection Agency revealed that about 2.5 tons of mercury were added to the environment during 1971 from rotary sewage flow distributors at municipal sewage treatment plants. The discharge of effluent water and sewage sludge into the sea results in an increase in the Hg level of the aquatic environment. The incineration of sewage results in the release of mercury, mainly in alkyl form, into the atmosphere and is detectable at a distance of several kilometres from the plant (Soldano et al, 1975).

The mercury level in sewage water gives an indication of the mercury emission in urban areas. Some 97% of the mercury present in sewage water can be removed by purification (Larsson, 1970). About 350kg of mercury per year were discharged into a treatment plant in Sweden which served 600000 inhabitants (Hanson, 1971). Larsson (1970) has observed that sludge from some 30 towns in Sweden had a mercury content ranging from 0.8 to 40mg/kg dry weight. He observed high mercury levels in strongly urbanised areas with differentiated industries. The sludge obtained from one plant and deposited at sea has been estimated to release approximately 1100lb of mercury per year. The incineration of sewage sludge from Chicago's Stickney Sludge Dryer releases 570lb of mercury per year. The US Environmental Protection Agency reports that mercury may be discharged at the rate of about 2700lb per year in the effluent of the Whites Point District, Los Angeles Sanitation Sewage Treatment Plant outfall. This estimate is based on a 24 hour composite sample taken during the autumn of 1970.

g. AGRICULTURAL USE OF PESTICIDES

The use of the mercury-based pesticides, phenyl- and alkylmercury, together with fertilisers can contribute significantly to environmental pollution. Mercury fungicides used to treat seed grain which is flooded after planting (e.g. rice) can directly enter the aquatic environment. Seed dressing can introduce as much as 10g/ha of mercury to the soil (Gowen et al, 1976). Grains treated with mercury may be consumed by wild animals, which can accumulate significant amounts of mercury in their bodies. The drainage of storage vessels used for the dipping of bulbs, corn, logs, and lumber in mercurial pesticides can be a potential source of Hg in soil and waterways. The Environmental Protection Agency has estimated that 9550lb of mercury were applied to the rice-growing area in Louisiana during 1969. Raw logs and lumber are often treated with mercurial pesticides and are dressed prior to use. The shavings are used in the manufacture of paper, wall-board, or as fuel which, when burned, contributes mercury directly to the atmosphere. A possible source of mercury contamination of remote New England lakes and ponds has been suggested to be the use of mercury-based preservatives to treat felled timber. Large quantities of mercury-based organic and inorganic fungicides are used to control turf-grass disease (Gilmour & Miller, 1973) on turf, including golf greens and tees, fairways, parks, cemeteries, and lawns. As much as 15lb of mercurial fungicides are added per acre of turf in a single application. Booer (1944) showed that organic mercurials are bio-degraded to Hg°. This bio-degradation has been confirmed by Kimura & Miller (1964). Tonomura et al (1972) later identified an enzyme of a strain of ***Pseudomonas*** in soil. This catalysed the reduction of mercury in organic, as well as inorganic, mercurials to metallic mercury. Further work on such bio-degradation and demethylation has been carried out by several workers (Spangler & Spigarelli, 1973; Rogers, 1977; Lexmond et al, 1976). The run-off from the above-mentioned areas results in the discharge of mercury to waterways. Mercurial fungicides have been banned in Sweden since 1966, when their role in the ecological cycle of seed-eating birds and their predators was proved (Berg et al, 1966)

Table 3.2: Mercury in Power-Plant Coals

Country	Mean Hg Content	Range	Reference
UK	0.28	0.025 - 1.0	Sabbioni et al, 1983
Irish Republic	0.28	- - - - -	Sabbioni et al, 1983
West Germany (a)	0.42	0.1 - 1.0	Sabbioni et al, 1983
France (a)	0.56	0.5 - 1.4	Kautz et al, 1975
Belgium	0.24	0.028 - 0.19	Sabbioni et al, 1983
Denmark	0.6	- - - - -	Block & Dams, 1975
Sweden (b)	0.28	0.60 - 0.4	Sabbioni et al, 1983
Italy	0.28	- - - - -	Larsson, 1970
Luxembourg	0.28	- - - - -	Sabbioni et al, 1983
Netherlands	0.28	- - - - -	Sabbioni et al, 1983
USA (166 samples)	0.18	0.02 - 1.6	Ruch et al, 1976

a) power plants, b) pit coals

h. FOSSIL FUELS

i Coal. Although the concentration of mercury in fossil fuels is low, the total emission arising from their burning is enormous because of the vast amount of fuel burnt every year. When coal is burnt, about 90% of the mercury originally trapped in the organic matrices of the vegetation which had transformed into coal, is released into the atmosphere while the remaining 10% stays in the ash (Billings & Matson, 1972; Billings et al, 1973; Kalb & Baldeck, 1973; Klein & Russel, 1975). The released mercury may eventually become dispersed over wide areas in the atmosphere (Klein & Russel, 1973). Rain water washes mercury from the atmosphere into the soil. Organic compounds in soil and low-pH water retain mercury and transform a large amount into monomethylmercury which increases the levels in fish (Jernelov, 1980) and effects re-emission of mercury species.

Mean Values in Coals. The mean levels of mercury throughout the world range from 0.01 to 3.00ppm (Fleischer, 1970). For British coals, the range is 0.13 to 2.28ppm, with reported mean values of 0.97ppm (Airey, 1982) or 0.5ppm (Gibson, 1979). For American coals, it is 3.3ppm (with a range of 0.07 to 33ppm), and for Australian oil it is 0.10ppm (with a range of 0.03 to 0.40ppm) (Porrit & Swaine, 1976), discharging 5300lb of mercury into the air annually.

The average mercury content in the coals, mostly used in coal-fired power plants in some European countries, the USA, and other places (Sabbioni et al, 1983), are presented in table 3.2.

The total amount of mercury that would be released in 1990 from coal-fired power plants in the EEC countries of Europe would be about 675 tonnes (Sabbioni et al, 1983).

ii Oil and Gas

In addition to coal, petroleum, natural gas, and oil brines have all been reported to contain significant quantities of mercury. Crude oils from the Cymric oilfield in California contain between 1.9 and 21ppm of mercury and the natural gases from the field are saturated with mercury vapour. During transport through pipelines, this vapour combines with hydrogen sulphide from the "sour" gases of other oilfields and is precipitated in the pipelines (White et al, 1970).

The latter authors have observed that hundreds of tons of mercury separate from crude oil at the local pumping station. They also reported that the light petroleum fraction of the Abbott Mine of California contained 100ppm of mercury and the tarry petroleum fraction of the mine contained 500ppm. Tar from a Wilbar Spring District oilfield in California contained 1ppm of mercury. Brines from the Cymric oilfield were found to contain up to 0.2ppm of mercury (White et al, 1970). Hanson (1971) reports that Swedish fuel oil may contain 2-4ppm of mercury while gasoline contains 2-17ppm of mercury. On the other hand, the reports of Larsson (1970) indicate that Swedish fuel oils contain an average of 3ppm of mercury. Thus, it is apparent that, although the total amount of mercury discharged into the environment due to oil burning, waste gas, and the use of natural gas, fuel oil, and gasoline as a source of power and heat is still not clearly known, these processes do release large quantities of mercury into the environment.

i. CEMENT MANUFACTURE

High-temperature (1500°C) burning releases mercury from the ingredients (limestone, shale, coal, gypsum, and sand). The mercury concentrations in these substances, apart from coal, ranged from 0.03ppm (sand) to 0.46ppm (limestone) (Fleischer, 1970). Assuming that world cement production is 7×10^8 tonnes, the amount of mercury released from the ingredients (excluding the coal component) is 250 tonnes per year (Airey, 1982). Other estimates for the mercury release from cement industries are: 100 tonnes/year (Weiss et al, 1971; Goldberg, 1978) and 500 tonnes/year (Heindryckx et al, 1974).

j. CATALYSTS

The use of mercury compounds as catalysts in chemical industries may bring about potential environmental hazards as will be apparent from the incident at Minamata Bay in Japan where the discharge of spent mercury catalysts from the plastics industry led to severe mercury poisoning involving about 50 families. The Study Group on Mercury Hazards (1970) has reported that a vinylchloride plant in Sweden discharged about 300kg of mercury per year. Several tons of spent mercury catalysts were dumped into the sea off the Swedish East Coast prior to 1964 (Hanson, 1971). Kurland et al (1960) have observed a mercury concentration of 12.5ppm in the muds of Galveston Bay adjacent to a holding basin where spent mercury catalyst used in a vinylchloride factory in Texas was incinerated and released into the atmosphere. Mercuric oxide catalysts are also discharged into waterways by factories manufacturing dyes with sulphonated anthraquinone products as raw materials.

k. ELECTRICAL INDUSTRIES

Mercury is used in the making of fluorescent and high-intensity discharge lamps, rectifiers, oscillators, power control switches, hot cathode tubes (thyratron and phanotron), and pool cathode tubes for high-frequency applications in radar, radio welding, etc.

Mercury dry cell batteries are used in many instruments, e.g. Geiger-Müller counters, radios, digital computers and calculators, hearing aids, military hardware, scientific and communication devices. Some 50% of the electrical uses of mercury involve the making of primary and dry cell batteries.

The mercury compounds used in dry cell batteries are inorganic mercurials. The contents of these batteries have been described by Ducrey (1971) and Devis (1972). Metallic zinc is usually used for the anode of a mercury battery, while the cathode consists of mercuric oxide mixed with graphite. In discarded batteries, the inorganic mercury is transformed into methylmercury (Yamaguchi et al, 1983) via a process which is entirely different from the bio-transformation described earlier by Jensen and Jernelov (1969). Elemental mercury can be dissolved in water, with or without any electrolytic dissociation. Reaction with carbon black, amorphous carbon, tar, or pitch can produce methylmercury compounds. In the presence of chlorides, the compound produced will be converted into methylmercury chloride. Apart from acetylene carbide and acetaldehyde, the acetylene-carbon black also contains several impurities which can produce methylmercury when mixed with inorganic mercury (Yamaguchi et al, 1971).

Fluorescent lamps contain 50-100mg of mercury. Breakages during the disposal of used fluorescent lamps can result in local concentrations of mercury vapour of up to $7mg/m^3$ (Freeman & Oppold, 1972).

The metal is also found in power stations (using mercury vapour-powered turbines), in nuclear reactors (using mercury as a neutron shield due to its ability to absorb neutrons), and in meters for direct current measurement.

I. INCINERATORS

Mercury-containing materials such as wood, paper, garbage, batteries, and used containers are often incinerated and this will result in the dispersion of mercury into the atmosphere. The US Environmental Protection Agency (US-EPA, 1970) found an average of 2.4ppm of mercury in 11 samples of refuse collected from six different incinerators, while New York's solid wastes contained 4.8 to 9ppm, due to discarded batteries, lamps and electrical gadgets. The seven incinerators in New York account for emissions ranging from 10000 to 19000kg/year. Nine Canadian cities emitted

4.44 tons of mercury in 1970 due to the burning of 1.8 million tons of refuse having an average mercury content of 2.4ppm (McLaren, 1973). It has been reported that Chicago's Calcument Incinerator discharges about 5400lb of mercury per year.

REFERENCES

- Airey, D. (1982) - Science of the Total Environment, 25, 19-40
- Berg, W., Johnels, A.G., Strand, B.S., and Westermark, T. (1966) - Oikos, 17, 71-83
- Billings, C.E. and Matson, W.R. (1972) - Science, 176, 1232-3
- Billings, C.E., Sacco, A.M., Matson, W.R., Griffin, W.R., Coniglio, W.R., Harley, R.A. (1973) - Journal of the Air Pollution Control Association, 23, 773-7
- Block, C. and Dams, R. (1975) - Environmental Science and Technology, 9, 146-50
- Booer, J.R. (1944) - Annals of Applied Biology, 31, 340-50
- Bouveng, H.O. (1967) - Oikos, Supplement 9
- Copplestone, J.F. and McArthur, D.A. (1967) - British Journal of Industrial Medicine, 24, 77-80
- Devis, J. (1972) - Dry Cell, Mercury Battery, in Encyclopedia of Science and Technology, McGraw Hill, Kodansha, Tokyo, 178-9, 454-7
- D'Itri, F.M. (1973) - in Bioassay Techniques and Environmental Chemistry (Edited by G.E.Glass), Ann Arbor Science, Michigan, 3-70
- Ducrey, L. (1971) - in Encyclopedia of Occupational Health and Safety, Vol. 1, (A-K), International Labour Office, Geneva, 163-4
- FAO (1978) - Yearbook of Fishery Statistics, FAO/UN, Rome, 342
- Fimreite, N. (1970) - Environmental Pollution, 1, 119
- Fleischer, M. (1970) - US Geological Survey Professional Paper 170, 6-13
- Forbes, G. and White, J. (1952) - British Medical Journal, 1, 899-901
- Freeman, T. and Oppold, J.A. (1972) - American Industrial Hygiene Journal, 33, 419-22
- Frykholm, K.O. (1970) - Svensk Tandlazkare Tidskrift, 63, 763-72
- Gardner, D. (1975) - Marine Pollution Bulletin, 6, 43-6

- Gibson, J. (1979) - Coal and the Environment, National Coal Board, London

- Gilmour, J.T. and Miller, M.S. (1973) - Journal of Environmental Quality, 3, 145-8

- Goldberg, E.D. (1978) - in Advances in Oceanography, Joint Oceanographic Assembly (Edited by H.Charmoch & G.Deacon), Scotland, 49-69

- Gowen, J.A., Wiersma, G.B., and Tai, H. (1976) - Pesticides Monitoring Journal, 10, 111-3

- Hanson, A. (1971) - 'Man-made Sources of Mercury' in the Royal Society of Canada Proceedings, Special Symposium on Mercury in Man's Environment, Ottawa, Canada, 201

- Heindryckx, R., Demuynck, M., Dams, R., Janssens, M., and Ralm, K.A. (1974) - Proceedings of an International Symposium on the Problems of Contamination of Man and his Environment by Mercury and Cadmium, Luxembourg, 3-5th July, 1973, CEC, Luxembourg, 135

- Iyer, K., Goodgold, J., Eberstein, A., and Berg, P. (1976) - Archives of Neurology, 33, 788-90

- Jensen, S. and Jernelov, A. (1969) - Nature, 223, 753

- Jernelov, A. (1980) - Environmental Science Research, 17, 211-22

- Joensuu, O.I. (1971) - Science 172, 1027-8

- Kalb, G.W. and Baldeck, C. (1973) - in Report EPA Contract No. 68-02-0225 Tra-Det Laboratories, Columbia, Ohio

- Katz, S. (1963) - Biochimica et Biophysica Acta, 68, 240

- Kautz, K., Kirsch, H., and Laufhütte, D.W. (1975) - VGB Kraftwerkstechnik, 55, 672-6

- Kimura, Y. and Miller, V.L. (1964) - Journal of Agricultural and Food Chemistry, 12, 253-7

- Klein, D.H. (1971) - in Mercury in the Western Environment (Edited by D.R.Buhler), Corvallis, Oregon, 3-38

- Klein, D.H., Andren, A.W., Carter, J.A., Emery, Z.F., Feldman, C., Fulkerson, W., Lyon, W.A., Ogle, J.C., Talmi, Y., Hook, R.I., and Bolton, N. (1975) - Environmental Science and Technology, 9, 973-9. By Hg and Cd, 279 (Proc. of an Int. Symp. Luxembourg, 3-5th July, 1973). Commission of the European Communities, Luxembourg

- Klein, D.H. and Russel, P. (1973) - Environmental Science and Technology, 7, 357-8

- Korringa, P. and Hagel, P. (1973) - Problems of the Contamination of Man and his Environment (Proceedings of an International Symposium, Luxenburg, 1973), Commission of the European Communities, Luxemburg, p.279

- Kothny, E.L. (1973) - in Trace Elements in the Environment (Edited by R.T.Gould), Advances in Chemistry Series, 123, 48-91

- Kurland, L.T., Faro, S.N., and Siedler, H. (1960) - World Neurology, 1, 370

- Larsson, J.E. (1970) - Environmental Mercury Research in Sweden, Swedish Environmental Protection Board Research Secretariat

- Lexmond, T.M., De Hann, F.A.M., and Frissel, M.J. (1976) - Netherlands Journal of Agricultural Science, 79-97

- McLaren, J.F. Ltd (1973) - National Inventory of Sources and Emission of Mercury, 1970, Report APCD 73-6, Environment of Canada, Ottawa, Canada

- Mercury Pesticides Registration Review Panel (1971) - A report on Mercury Hazards in the Environment

- Miller, D.R. and Buchanan, J.M. (1979) - Atmospheric Transport of Mercury: Exposure Commitment and Uncertainty Calculations. A Technical Report MARC No. 14, Chelsea College, London

- National Materials Advisory Board (1970) - Trends in the Usage of Mercury, Report on the Panel on Mercury. Journal of Metals (May), 28

- Neal, P.A., Flinn, R.H., Edwards, T.I., Renhart, W.H., Dallavalle, J.M., Goldman, F.H., Armstrong, D.W., Gray, A.S., Coleman, A.L., and Postman, B.F. (1941) - Mercurialism and Its Control in the Felt-Hat Industry. US Public Health Bulletin No.263

- Porritt, R.E. and Swaine, D.J. (1976) - Institute of Fuel Sydney Conference, Paper 18, Special Committee to Study Health and Environment

- Rogers, R.D. (1970) - Journal of Environmental Quality, 5, 454-8

- Rogers, R.D. (1977) - US Environmental Protection Agency, EPH-600/3-77007

- Ruch, R.R., Cahill, R.A., Frost, J.K., Camp, L.R., and Gluskoter, H.J. (1976) - Proceedings of the International Conference on Modern Trends in Activation Analysis, Munich, 1042-8

- Ruch, R.R., Gluskoter, H.I., and Kennedy, E.I. (1971) - Environmental Geology Notes, No.43, 1-14

- Rupp, N.W. and Paffenbarger, G.C. (1971) - Journal of the American Dental Association, 82, 401

- Sabbioni, E.L., Götz, A., Springer, A., and Pietra, R. (1983) - Science of the Total Environment, 29, 213-27

- Slemr, F., Seiler, W., Eberling, C., and Roggendorf, P. (1979) - Analytica Chimica Acta, 110, 35-47

- Slemr, F., Seiler, W., and Schuster, G. (1981) - Journal of Geophysical Research, 86, 1159-66

- Smith, R.G., Vorwald, A.J., Patil, L.S., and Mooney, T.F. (1970) - American Industrial Hygiene Association Journal, 31, 687-700

- Soldano, B.A., Bien, P, and Kwan, P. (1975) - Atmospheric Environment, 9, 941-4

- Soremark, R. (1968) - Journal of Prosthetic Dentistry, 20, 531

- Spangler, W.J. and Spigarelli, J.L. (1973) - Science, 180, 192-3

- Special Committee to study Health and Environmental Effect of Increased Coal Production. Basis for a National Reclamation Plan Environment Health Perspective (1979), 33, 159-76

- Stahl, O.R. (1969) - Preliminary Air Pollution Survey of Mercury and Its Compounds, a Literature Review, US Department of Health and Welfare, Public Health Service, National Air Pollution Control Administration

- Study Group on Mercury Hazards (1970) - Hazards of Mercury, Special Report to the Secretary's Pesticide Advisory Committee, Department of Health, Education and Welfare, and Environmental Protection Agency

- Tonomura, K., Eurukowa, K., and Yamada, M. (1972) - in Environmental Toxicology of Pesticides (Edited by F.Matsumura, G.B.Boush, & T.Masato), Academic Press Incorporated, New York, 115-33

- US Environmental Protection Agency (1975) - Materials Balance and Technology Assessment of Mercury and its Compounds on National and Regional Bases, October, 1975, Final Report. Prepared for Environmental Protection Agency, Office of Toxic Substances, by URS Research Company, EPA-560/3-75-007, US Environmental Protection Agency, Washington DC

- US Geological Survey (1968) - Minerals Year Book

- Weiss, H.V., Koide, M., and Goldberg, E.D. (1971) - Science, 174, 692-4

- West, J.M. (1970) - Mercury in 1969. US Bureau of Mines Minerals Year Book, 1

- Westerstrom, L.W. and Harris, R.E. (1977) - Coal Bituminous and Lignite, in Minerals Year Book, Vol. 1, US Government Printing Office, Washington DC, 87-456

Sources of Mercury in the Environment

- White, D.E., Hinkle, M.E., and Barnes, I. (1970) - in Mercury in the Environment, US Geological Survey Professional Paper 713, 25-8

- WHO (1976) - Environmental Health Criteria Mercury, World Health Organisation, Geneva 1-131

- Yamaguchi, S., Matsumoto, H., Hoshide, M., Matsuo, S., and Kaku, S. (1971) - Archives of Environmental Health, 23, 196-201

- Yamaguchi, S., Shimogo, N., Sano, K., and Fujishima, T. (1983) - Science of the Total Environment, 27, 53-8

CHAPTER IV

MOVEMENT OF MERCURY IN THE ENVIRONMENT

INTRODUCTION

Because of its high volatility, mercury becomes dispersed over a very large fraction of the atmosphere (Goldberg, 1976). It is dispersed as vapour or as particles associated with dust, smoke, volcanic gases, and the natural degassing of soils. Some dust particles may become entrained in high tropospheric and stratospheric wind currents (Crockett & Kinnison, 1977; Goldberg, 1976; Windom et al, 1975). Any mercury deposited back into the soil may be revolatilised in aerobic terrestrial environments (Rogers, 1977; Miller & Buchanan, 1979).

The high volatility of the metallic phase and of inorganic (mainly mercuric chloride) and organic compounds (mainly monomethyl- or dimethylmercury) leads to wide-ranging transport of mercury in air. Annually, more than 8000 tonnes of mercury are mined and processed for use in industry (Drake, 1977). Mercury is released into the atmosphere when fossil fuels are burnt and ores are roasted (Weiss et al. 1971; Airey, 1982).

Besides chemical and other industrial activities, agriculture and mining contribute major amounts of mercury to the ecosystem, including land (soil), water, and air. Via bacterial action, a portion of this would be converted to methylmercury (Friberg & Vostal, 1972) and some would be concentrated in fish (Cole, 1979). Fish-eaters then accumulate mercury in their bodies.

The amount of mercury in air varies with height but is often 10 to 20 times greater at ground level than at 120m (Williston, 1968).

The mercury released can either stay close to the source for long periods (Gardner & Riley, 1973; Airey & Jones, 1982), or be dispersed world-wide within several weeks

(Gardner 1975; Windom et al, 1975). The sea-water, air, and also the hair of the people, of the northern hemisphere are found to contain greater amounts of mercury than those in the southern hemisphere. This is because of the greater industrialisation in the north, and the characteristic stratospheric circulation system, which leads to the redeposition of pollutants from the mid-latitude industrial northern hemisphere within the same hemisphere.

Via global recycling, mercury finally accumulates in the sea. The ratio of the natural input to the anthropogenic input is about 4:1 (Miller & Buchanan, 1979). Annually, man consumes about 15000kg of mercury by eating fish (world fish catch: 73.5 million tonnes, x average Hg content: 0.2ppm; FAO, 1978).

The sea serves as the ultimate sink for mercury. However, small amounts of mercury may be released from sea and aquatic systems by bio-methylation and volatilisation into the air and finally become fixed in aquatic sediments. There are indications that mercury levels in the sediments deposited in an English lake have increased since the industrial revolution (Aston et al, 1973).

Williston (1964) indicated that the presence of a water table above mercury deposits does not greatly reduce the rate of mercury loss by vaporization. He found 0.6 to 0.7mg/m^{3}Hg some twenty miles off-shore over the Pacific ocean. This suggested that the land surface is the principal source of mercury in the atmosphere.

Little information is available as to the extent of the reactions of gaseous mercury with earth materials. Noble metals such as platinum, gold, and silver readily form amalgams with mercury. Organic matter and clays absorb gaseous mercury and therefore the atmospheric mercury level is continually being decreased by reaction with air-borne particulate matter and land surfaces. Some of the mercury bound up in wet soil can be revolatilised as the soil dries, while some may be trapped by humus material.

MEDIA

A. AIR

(a) Atmosphere. In air, 50% of the mercury content is elemental (Johnson & Braman, 1974) and 4% is particulate ore.

The saturation level of mercury in air in equilibrium with metallic mercury is known to increase exponentially with temperature (Vaughn, 1967). At 20°C, the normal equilibrium content of mercury in air is about 13.2ng/cm^3, whereas the background content is 2 to 8ng/m^3 (Williston, 1968). Jepson (1973) measured atmospheric mercury concentrations in many industrial, rural, residential, and aquatic areas and found background levels of 0 to $800ng/m^3$, with maximum levels of up to $28000ng/m^3$. High levels are found over industrial areas. The levels found are:

Belgium	6.7 to $7.4ng/m^3$ (Dumarey et al, 1979)
England	1 to $6ng/m^3$ (Airey, 1982)
West Germany	1 to $4ng/m^3$ (Slemr et al, 1979)

The mean value of the Hg content of Liverpool air is $18.9ng/m^3$ (Airey, 1982). The conditions of soil temperature, mercury species, pH, redox conditions, humus content, and plant cover can affect the release of mercury (Anderssen, 1967; Rogers, 1977 Kozuchowski & Johnson, 1978; Hogstrom et al, 1979). Mercury becomes globally dispersed throughout the atmosphere and its residence time there varies from 5.5 to 90 days. However, the content decreases with altitude according to the relationship:

$$C_z = C_o e^{-kz}$$

where C_o is the ground concentration, C_z is the concentration at an altitude, z, and k is a constant of the order of 1mm. Studies on ice cores from Greenland suggested that there had been no systematic change in the mercury concentration in the atmosphere over the last 150 years. Greenland, polar regions, and open mid-oceanic atmospheres exhibit little variation, indicating that anthropogenic sources contribute to the continental landmass areas, where again there is always a drift from east to west.

The mercury contents in air, reported by various workers, are presented in table 4.1.

Anomalous mercury contents in air are found over mercury ore deposits and over volcanoes (Siegel et al, 1975) followed by sulphide deposits and recently faulted zones. The total emission from land degassing (1.78×10^{10}g/yr), volcanoes (2×10^7g/yr), and anthropogenic sources (1.00×10^{10}g/yr) contribute to the atmosphere and to slower feedback recycling.

Large quantities of mercury are released into the atmosphere from anthropogenic sources such as the manufacture of chlorine, the industrial use of catalysts, the mining and refining of ores, mining and burning of fossil fuels, incineration of trash (including paper, wood, containers etc), processing of phosphate rocks, etc.

In sludge digesters, methylated mercury is produced but is released to the atmosphere as mercury vapour when the gas is burned as fuel or waste flares.

Air moving over highly polluted Hillsborough Bay at Tampa, Florida was found to contain a five times higher total mercury compound content as compared to the average content in the air moving over land (Johnson & Braman, 1974).

Table 4.1: Mercury Content in Air

Average(ng/m^3)	Sampling Site	Reference
20		Saukov (1953)
0.5 - 50	San Francisco Bay	Williston (1968)
1.6 - 7.2	south-west USA	McCarthy et al (1969)
1 - 10	various	Fleischer (1970)
1.9	Michigan & Indiana (unpolluted)	Dams et al (1970)
4.8	Michigan & Indiana (polluted)	Dams et al (1970)
3 - 9	various locations in USA	McCarthy et al (1970)
1.4 - 15.5	various locations in Sweden	Ferm & Larsson (1973)
0.17	Heidelberg	Bogen (1973)
3.2	Tampa area of Florida	Johnson & Braman (1974)
2 - 10	several locations	Johansson & Boyle (1971)
30 - 1600	over mercury deposits	Johansson & Boyle (1971)

A considerable amount of mercury is released into the atmosphere from soil and mineralised land areas and by volcanic activity. The rate of vaporization of mercury and its compounds follows the pattern:

$$Hg > Hg_2Cl_2 > HgCl_2 > HgS > HgO$$

(Kokshoy & Bradshaw, 1969). The vapour pressure of the methyl and ethyl mercurials (0.0008 to 0.023mm Hg at 25°C) is greater than that of the phenyl form (1.7×10^{-5} to 0.08×10^{-4} mm Hg at 35°C, Phillips et al, 1959). Of almost all the mercury compounds, methylmercury chloride is the most volatile.

The vaporization rate of mercury approximately doubles for every 10°C increase

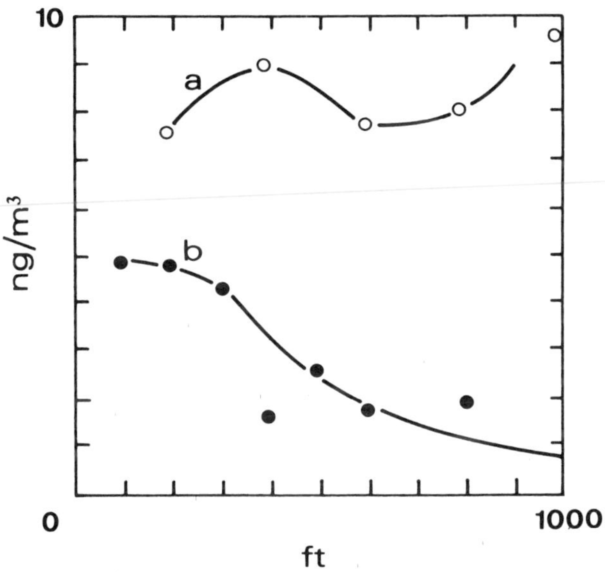

Figure 4.1 Mercury in the Atmosphere as a Function of Altitude
(a: 27th April 1968, b: 1st January 1968. From McCarthy et al, 1969)

in temperature (Stahl, 1969). Similarly, the saturation level of mercury in air in equilibrium with metallic mercury increases logarithmically with increasing temperature (Vaughn, 1967). These characteristics may explain the correlation of the atmospheric mercury level with the season and the time of day. For example, the curve for January (figure 4.1) indicates that, above 300ft, the mercury concentration dropped markedly, whereas data collected at the same site in late April exhibited no obvious trend. Figure 4.1 also demonstrates that lower mercury levels are observed in January than in April. Williston (1968) found similar mercury contents in air in the San Francisco Bay area.

At Stromsardt in central Sweden, during 1969 and 1970, the monthly total mercury contents in precipitation samples exhibited seasonal changes related to evaporation during the summer months (Study Group on Mercury Hazards, 1970). A low value of 0.08 μg/l of mercury was measured in the samples during December, while a high value of 1.45 μg/l was measured in July.

Table 4.2: Maximum Mercury Concentrations in Air Measured at Scattered Mineral and Non-mineralised Areas of the Western United States (US Geological Survey Professional Paper 713, 1970)

Sample Location	0ft (ng/m^3)	400ft (ng/m^3)
Mercury Mines		
Ord mine, Mazatzal mountains, Arizona	20000	108
Silver Cloud mine, Battle mountain, Nevada	2000	24
Dome Rock mountains, Arizona	128	57
Base and Precious Metal Mines		
Cerro Colorado, Arizona	1500	24
Cortez gold mine, Crescent valley, Nevada	180	55
Coeur d'Alene mining district	68	-
San Zavier, Arizona	-	25
Porphyry Copper Mines		
Silver Bell mine, Arizona	-	53
Esperanza mine, Arizona	-	32
Vekol mountains, Arizona	-	32
Ajo mine, Arizona	-	30
Mission mine, Arizona	-	24
Twin Buttes mine, Arizona	-	22
Puma mine, Arizona	-	18
Safford, Arizona	-	7
Non-mineralised Areas		
Blythe, California	-	9
Gila Bend, California	-	4
Salton Sea, California	-	3.5
Arvaca, Arizona	-	3

Table 4.2 shows the mercury concentration in air over 15 ore deposits and above 4 non-mineralised areas. In general, the concentration of mercury is greatest over

mercury mines, lower over base and precious metal mines, and lowest over porphyry copper mines. The mercury concentration in air over non-mineralised areas ranged from 3 to 9ng/m^3. The reports of Neville (1967) indicate that the mercury vapour concentration is between 10^5 and 2×10^6ng/m^3 in the mercury mine at Idria, Yugoslavia; the concentration in the mercury processing plant being 6×10^4 to 9.7×10^5ng/m^3. Sergeev (1961) has collected soil air from a bore-hole between 1 and 2m deep and found mercury vapour concentrations of between 0 and 100ng/m^3, whereas those collected 1m above the surface contained between 10 and 20ng/m^3. Besides seasonal variations in the mercury content of air, there are daily variations. The maximum amount of mercury in air was found at mid-day. Much smaller amounts were found in the morning and in the evening. The daily variations are due to changes in barometric pressure. When the pressure begins to fall, mercury is released into the atmosphere and reaches a maximum when the rate of fall of the pressure is greatest. The effect of temperature is less obvious and the maximum daily temperature commonly occurs some 2 to 4h later than the time at which the maximum mercury level is found in air (McCarthy et al, 1970).

Williston (1968) suggested that, because of the ability of mercury to absorb on any and all surfaces, dust particles in the air will normally carry more than the expected amount of mercury, and mercury evaporated into the air will tend to concentrate on dust particles. He observed that high atmospheric mercury levels always coincided with high smog levels in the San Francisco Bay area. He decided that it was mere speculation whether the high mercury content of smog was coincidental or whether the catalytic effect of mercury vapour under the influence of ultra-violet light contributed to the smog.

Annually, more than 8000 tonnes of mercury ore are mined and processed for industry (Drake, 1977). The fumes from various industrial and smelting processes contain gaseous and particulate mercury. It is also released by the burning of fossil fuels and ores (Weiss et al, 1971; Airey, 1982). The dust released by the mining of sulphide-bearing minerals may be a significant source of mercury. The saturation point of mercury in air in equilibrium with metallic mercury increases logarithmically with increasing temperature (Vaughn, 1967). The mercury content of soil air over a mercury ore deposit was observed to be 100ng/m^3, whereas the atmospheric air immediately over the deposit gave a value of 10 to 20ng/m^3. This undersaturation, by a factor of 10^4, of the air over the mercury deposit (the saturation value of mercury in air at

17°C being 10^6 ng/m^3) indicates a rapid exchange of soil air with atmospheric air in comparison to the rate of evaporation of mercury. McCarthy et al (1969) suggested that mercury in soil air samples was unrelated to the concentration of mercury in soil. Therefore, a large part of the mercury appeared to come from a greater depth.

Table 4.3: Concentration (ng/SCM) of Particulate Mercury in the Atmosphere

Location	Collection Medium	Mean	Range	Ref.
Tropical North Atlantic	Whatman 41	0.065*	- - - -	a
Atlantic	Whatman 41	-	<0.02-2.25	b
Switzerland	Whatman 41	0.030	<0.04-0.90	c
South Atlantic Bight	glass fibre	0.80	0.30-3.80	d
Long Island Sound	glass fibre	0.29	0.01-1.20	e
Pigeon Key, Florida	glass fibre	0.03	0.01-0.05	e
Tampa, Florida	glass fibre	0.40	0.0-1.40	f
O.S. Lima (57°N)	Millipore	0.41	0.13-0.089	g
North Atlantic (mid. lat.)	Millipore	0.013	0.0-0.017	g
South Atlantic (mid./high lat.)	Millipore	0.008	0.002-0.021	g

* geometric mean

a: Buat-Menard & Chesselet (1979)
b: Arnold (1977)
c: Dams & De Jonge (1976)
d: Windom & Taylor (1979)
e: Fitzgerald & Gill (1979)
f: Johnson & Braman (1974)
g: Millard & Griffin (1980)

(b) Particulate mercury. Determinations of the concentrations and distributions of particulate trace elements in the marine atmosphere (Chester et al, 1974; National Academy of Sciences, 1978; Buat-Menard & Chesselet, 1979; Rahn, 1980) have become of great importance in understanding global geochemical mass balances (Mackenzie & Wollast, 1977; Lantzy & Mackenzie, 1979). The concentration of particulate Hg in the marine atmosphere appears to be substantially lower than that of the vapour phase (Fitzgerald & Gill, 1979).

Table 4.4: Mercury Content in Atmospheric Precipitation

Area	Mean (ng/l)	Range (ng/l)	Ref.
Rainwater			
Sweden	140	0 - 480	a
Sweden	100	0 - 53	b
Sweden	330	0 - 200	c,d
Southern New England, USA	10	5 - 15	e
Northern New England, USA	60	- - -	f
Pacific Ocean	1	- - -	g
Tutuila Island, American Samoa	0.8	0.6 - 1.0	e
North Pacific Ocean	-	10 - 50	h
North Sea	30	17 - 58	i
Snow and Ice			
Greenland ice sheet (800BC to AD1952)	60	43 - 77	j
Greenland ice sheet (AD1952 - 1965)	125	87 - 230	j
Greenland ice sheet (1971)	-	29 - 78	k
Greenland ice sheet (1850)	13	3 - 23	l
Greenland ice sheet (1870)	169	154 - 184	l
Greenland ice sheet (1930)	100	88 - 112	l
Greenland ice sheet (1940)	50	40 - 60	l
Greenland ice sheet (1940)	10	- - -	m
Sweden	-	80 - 500	n
Sweden (unpolluted area)	-	80 - 120	o
Sweden (over chlor-alkali plants)	~1100	- - -	o
Sweden (unpolluted area)	-	10 - 70	p

a: Stock & Cucuel (1934)
b: Sevel (1975)
c: Eriksson (1967)
d: Brune (1969)
e: Fogg & Fitzgerald (1979)
f: Schlesinger et al (1974)
g: Matsunaga & Goto (1976)
h: Nishimura (1979)
i: Cambray et al (1979)
j: Weiss et al (1971)
k: Weiss et al (1975)
l: Carr & Wilkniss (1973)
m: Appelquist et al (1978)
n: Stråby (1968)
o: Jernelöv & Wallin (1973)
p: Ferm & Larsson (1973)

Atmospheric particulate matter is enriched in Hg relative to normal crustal material or sea-salt nuclei. Anthropogenic inputs may be responsible for the observed enrichment of Hg-rich particulates. Table 4.3 shows recent determinations of mercury in atmospheric particulates.
The upper limits for the flux of Hg from the atmosphere to the North Atlantic and South Atlantic were calculated to be between 380 and 600 tonnes/yr and between 100 and 150 tonnes/yr, respectively (Millard & Griffin, 1980).

(c) Rain wash. Some of the airborne particulate mercury returns to the surface by dry fall-out, and some is washed down by rainfall. However, rain does not completely wash mercury out of the atmosphere (Stock & Cucuel, 1934; Fogg & Fitzgerald, 1979). Nevertheless, McCarthy et al (1970) could not detect mercury in air after a rainstorm although the concentration was $20ng/m^3$ Hg in the air near a mine one day before the storm. Kothny (1973) suggested that the decrease in mercury levels observed after storms might also be due to increased ventilation at meteorological fronts, especially cold fronts.
Wiklander (1965) calculated a yield of 1.2g/ha-yr of atmospheric mercury via rainfall in Sweden. In industrial zones, more mercury is precipitated by dry fall-out than by rainfall during dry seasons. Indeed, Dams et al (1970) found a two and a half times higher value of particulate mercury in the atmosphere of an industrial area of Chicago as compared to a rural area.
The levels of atmospheric mercury measured in Liverpool during June, 1972 and March, 1974 are given by Airey (1982, p31). Low atmospheric mercury concentrations occurred for long periods after rain and the levels did not increase until the ground became dry. The mean value was 16.7 (range: $6 - 30ng/m^3$).

(d) Snow and ice. The mercury contents in the permanent snow-fields of Greenland indicate that snow removes mercury from the atmosphere (Weiss et al, 1971). The mercury levels in rain and snow from various locations are presented in table 4.4.
It is now fairly well established that deposition of mercury in ice in the polar region has not changed significantly since 800BC (Brouzes et al, 1977).
Deposition of mercury onto the land is estimated to amount to 2.5×10^{11} to 1.5×10^{11} g/yr (Weiss et al, 1971), although the fractions due to different vehicles of deposition are not yet known. The fall-out of mercury in vapour form is much slower,

because raindrops (pH less than 7.0), running water, and moisture are not good sinks for mercury (Braman & Johnson, 1974).

B. SOIL

Mercury enters soils via the disposal of sewage sludge, dry fall-out from the atmosphere, rainfall, and the agricultural use of mercury-based pesticides.

Mercury is strongly absorbed by soils and sediments and is desorbed very slowly. That mercury is fixed also becomes evident from the fact that some soil horizons over mercury deposits are rich in mercury in spite of the high vapour pressure of free mercury and the solubility of the chlorides of mercury. Ross and Stewart (1962) confirmed the affinity of certain soils for mercury when they observed that mercury sprays failed to migrate more than two inches from the surface, even after several years.

Isomorphous substitution for barium (Ba^{2+} radius = 0.134nm) and, to a lesser extent, for calcium (Ca^{2+} radius = 0.099nm) may slowly remove mercury from water and soil solutions. The much greater electronegativity of mercury compared to that of calcium, and the smaller ionic radius of divalent mercury compared to that of barium, tend to limit its solid solubility for barium and calcium. Nevertheless, in districts containing metallic mercury, barites may contain 20 to 190ppm mercury (Saukov, 1946).

Very little is known of the cation exchange properties of mercury. Ginzburg (1960) has stated that divalent ions form the following series with regard to their uptake by montmorillonite from aqueous solutions:

$$Pb > Cu > Ca > Ba > Mg > Hg$$

and that the order of replacement is:

$$Mg > Ba > Ca > Cu > Pb$$

The energies of adsorption by kaolinite are such that:

$$Hg > Cu > Pb$$

and the calcium replacement propensity is:

$$Pb > Cu > Hg$$

Experimental data indicate that mercury is absorbed slowly by membrane filters whereas peat moss, microcrystalline oxides, and soils absorb rapidly. Cellulose acetate membrane filters (2in, 0.45) absorbed 1.7% of the mercury in 50ml solution in a single pass while soils, peat moss and microcrystalline manganese oxides absorbed 50 to 100% of the mercury in 0.5g samples within 1h.

About 10 to 20% and 30 to 40% mercury, respectively, were desorbed from the manganese oxide samples in 1h by filtered tap water and by saline solution (0.5N). Subsequent desorption in the saline solution released small amounts of mercury.

Thus, mercury in solution is rapidly taken up by microcrystalline oxides, peat moss and soils, and absorption is irreversible. However, it is not clear whether uptake by these earth samples is due to absorption of cationic mercury or is due to reduction to metallic mercury, which is the stable form.

In soils, the clay minerals, sesquioxides and organic materials (humus) are responsible for the retention of elements. Mercury can be retained as a result of ion exchange, or chemisorption in the case of sesquioxides, where hydroxoligands and hydroxobridges are essential for retention (Forbes et al, 1974). At a pH of about 6, the adsorption of mercury by clay minerals is at a maximum.

The fine mineral constituents, particularly sulphides and organic mercurials, are the other sources of the mercury content of soil. The following sequence for the retention of inorganic mercury under neutral conditions has been determined (Anderssen, 1970):

$Al(OH)_3$ < Kaolinite < montmorillonite < illitic clay soil < laterite soil < organic soils < $Fe_2O_3nH_2O$. In acidic situations in podsols, organic matter predominantly absorbs mercury whereas, in neutral soils, iron oxides and clay minerals dominate Hg absorption. In soils with low organic matter contents, acidic conditions will increase Hg mobility (Trost & Bisque, 1972). Removal of Hg by leaching is more likely in acidic soils and evaporation as Hg^o is more likely in alkaline and neutral soils.

As discussed in chapter 2, mineralised deposits and tectonic zones are responsible for anomalous values found in the soils of certain regions. Atmospheric precipitation may bring down mercury dispersed from very remote sources.

The soil air over known mercury deposits exhibits well-defined anomalous values of between 100 and $5000ng/m^3$ Hg (Furzov, 1970; Karasik & Bol'shakov, 1965;

Khayretdinov, 1971; McCarthy, 1972; Meyer & Evans, 1973), whereas the background value is about $10ng/m^3$. Over polymetallic sulphide and precious metal deposits, the soil air samples also show anomalous values which define the mineralised zones (McNerney & Buseck, 1973; McCarthy et al, 1970; Meyer & Evans, 1973: Hawkes & Williston, 1962; Stevens et al, 1969).

In neutral soils, the mineral constituents and humus retain mercury but, in acid environments, the Hg is released from minerals, and the simultaneous microbial decomposition of organic matter releases mercury into the soil gas. Humic acid reduces mercuric mercury to Hg° and thereby releases it to the atmosphere (Alberts et al, 1974).

Inorganic mercury complexes are commonly retained on the top-soil with organic and mineral constituents. Vaporization becomes increasingly important in soils as the mercury content increases, and is mediated by microbial and non-microbial processes.

Fertilisers and fungicides contribute significantly to the Hg content of the soil. Mercury can occur in three valence states, Hg°, Hg_2^{2+}, and Hg^{2+}, depending upon the redox conditions of the soil. However, these are greatly modified by the microbial activity and are also a function of the Cl^- concentration and pH of the solution. In general, Hg° is considerably more available than Hg^{2+} because of the lipid solubility of the former (Hughes, 1957).

The thermodynamic data (Anfält et al, 1968) and the stability diagram of inorganic mercury species as a function of pH and redox potential (Anderssen, 1970), are available.

Microbial and biochemical reactions in soils and sediments lead to the formation of methylmercury and the more volatile dimethylmercury (Jensen & Jernelov, 1969). A strain of ***Clostridium***, an anaerobe isolated from soil, may be responsible for the methylation of mercury (Yamada & Tonomura, 1972).

Abiotic methylation in association with humic/fulvic acid was also demonstrated experimentally by Rogers (1977). Both biological and abiological methylation seem to be possible in soils. After formation, dimethylmercury evaporates rapidly through dry and porous soils, whereas an increasing moisture content retards evaporation (Frissel et al, 1971). Since both methyl- and dimethylmercury are found in gas from normal soil, they may play a role, in the mercury turn-over, which is similar to that played by Hg° vapour. In near-ground air samples, Johnson and Braman (1974) found some 25% of organic mercurials and 75% of inorganic species.

The mercury distribution in size-separated aerosols collected using an eight-stage Anderson sampler at Bhabha Atomic Research Centre*, Trombay (Bombay), and analysed by NAA, exhibits two distinct peaks at 0.5 and 6μm (Sadasivan, 1981). This suggests two sources of mercury. The bimodal distribution is in agreement with earlier observations (Rahn, 1976). The sources of mercury are a coal-fired power plant and caustic soda and chlorine plants located north and east of the sampling site.

C. WATER

Upon discharge into an aquatic environment, mercury, like any other metal, becomes partitioned between the solid (particulate) and liquid (solution) phases. Within each phase, further partitioning or speciation occurs among specific ligands to an extent which is determined by the ligand concentration and the strength of each metal-ligand association. Mercury, because it exchanges slowly, accumulates in organisms more efficiently than do other elements (e.g. cobalt) which exchange rapidly (Luoma, 1983).

Significant quantities of mercury are discharged to waterways from sources such as chloride manufacture, pulp and paper treatment, laboratories, hospitals, dental clinics, the manufacture and use of paint, the industrial use of catalysts, the mining and refining of ores, the disposal of used mercury containers, sewage treatment, and the use of mercury-based pesticides.

The erosion caused by rainfall and leaching transport a fraction of the atmospheric mercury from land surfaces to streams and other waterways. In addition to this transport, a fraction of the atmospheric mercury is returned to bodies of water by dry fall-out and rainfall.

According to Aidin'yan and Belavskaya (1963), appreciable amounts of mercury go into solution when ground water reacts with cinnabar or other mercury minerals, but this is removed almost completely when the solution is passed over mudstones. However, mercury is removed to a great extent by adsorption on $Fe(OH)_3$ or clay from sea-water (Dall'Aglio, 1968). Via chemical weathering alone, about 230 tons of mercury ore are leached out per year (Joensüü, 1971).

*The Trombay area, including BARC, is surrounded by refineries, petrochemical and fertiliser industries and, further away in the city of Bombay, are various industries such as chemicals (paints, etc), and automobile and iron works. There is a small incinerator some 12km west of the sampling site.

Table 4.5: Mercury Levels in Natural Waters outside of the United States

Location	Concentration (ppb)	Reference
sea-water, vicinity of Heligoland	0.03	a
Lamapar Deep	0.08 - 0.15	b
Ramapo Deep (south-east of Honshu)	0.15 - 0.27	c
Minamata Bay	1.6 - 3.6	d
sea-water, USSR	0.7 - 2	e
Volga, Don, Araks, & Danube rivers	1 - 2.0	f
rivers, European USSR	0.4 - 2.8	g
Armenian rivers & Swan Lake	1 - 3	h
rivers near Abkhasia Hg deposits, USSR	0.5 - 3.6	i
natural waters of Germany	0.01 - 0.05	a
Saale river, Germany	0.035 - 0.145	j
uncontaminated river waters, Italy	0.01 - 0.05	k
rivers near Hg deposits, Italy	up to 136	k

a: Stock & Cucuel (1934)	g: Aidin'yan (1962)
b: Hamaguchi et al (1967)	h: Aidin'yan (1963)
c: Hosohara (1961)	i: Zautashvili (1966)
d: Hosohara et al (1961)	j: Heide et al (1957)
e: Aidin'yan (1962)	k: Dall'Aglio (1968)
f: Aidin'yan & Belavskaya (1963)	

In water, mercurous chloride (Hg_2Cl_2) is only slightly soluble (0.002g/l), whereas mercuric chloride ($HgCl_2$) is highly soluble (69g/l at 25°C). Under aqueous conditions, mercurous chloride dissociates according to the reaction:

$$Hg^{2+} \rightleftharpoons Hg^{\circ} + Hg^{2+}$$

This reaction is promoted by ultraviolet radiation (Sidgwick, 1950). The insoluble basic sulfate, $Hg_2SO_4.HgO.H_2O$, is also likely to form by the oxidation of mercury-bearing sulphide ores (James, 1962). Due to its high solubility, $HgCl_2$ can be

leached out and reach streams via run-off, under-flow or ground-water discharge.
Both solute and particulate mercury may be released directly to the fluvial environment via streams which cut through mercury-bearing deposits. Thus, significant amounts of mercury may be added to streams by thermal springs, non-thermal springs, and mine drainage.
According to a survey of a total of 73 samples (US Geological Survey Professional Paper 713), 34 contained less than 0.1ppb, 27 contained between 0.1 and 1.0ppb, 10 samples ranged from 1.0 to 5.0ppb, and only 2 samples contained more than 5ppb, which is the US Public Health Service limit for potable water supplies. Except for the immediate vicinity of waste out-falls, the mercury concentration in surface water generally does not exceed tolerable limits.
The levels of mercury in surface water in other parts of the world usually fall within the same low range of values observed in the US (table 4.5).
Samples from rivers draining mercury deposits contain excessive amounts of mercury. Mercury-bearing minerals have been found in suspended particulate matter some 15 to 35km downstream from mercury ore deposits in the river Yaghob-Dar'ya.
Dall'Aglio (1968) observed a regular decrease in mercury content down the Paglia river in Italy. The mercury concentration in the stream decreased from 136ppb to 104ppb. Sediments from the Wisconsia river (USA) contained 560000ppb at a chemical company out-fall. However, at a distance of 4 miles downstream, the sediment contained 50000ppb (Chemical and Engineering News, 1970). Further downstream, at a distance of 21.4 miles, the mercury concentration decreased to 400ppb. Heide et al (1957) have found a progressive increase in the downstream mercury concentration in the Saale river (Germany), which is presumably due to industrial pollution.

D. SEDIMENTS

Physicochemical processes acting on the rocks and soils of a catchment area normally control the concentration of trace metals in river and estuary environments. The annual load of mercury in the Rhine estuary is about 53 tons (De Groot et al, 1973; Forstner & Wittmann, 1979). Dissolved mercury in water behaves conservatively in some estuaries (Windom & Taylor, 1979) and may not be involved in co-precipitation processes with Fe, Mn or organic material. Mercury levels in estuary sediments do not usually correlate with other geochemical variables (e.g. elements), suggesting that it is not actively associated with precipitation processes in the estuary. Field and

laboratory studies have shown that desorption of Hg from sediments takes place at higher ionic strengths (DeGroot, 1973; Frenet-Robin & Ottmann, 1978; Ramamoorty & Rust, 1978). A study of the distribution of mercury in the sediments of the Plym (Millward & Herbert, 1981) shows that it (18 - 2610ng/g) lacks any correlation with other sedimentary variables like Fe, Mn, clay content, or organic elements. Its distribution appears to be governed by sewage inputs rather than by drainage from a waste disposal site. A cluster analysis of 41 analysed samples indicated the presence of three geochemical facies. Cluster facies P was located upstream and had sediments that were poorly sorted, with Hg concentrations greater than those found in the catchment area. Cluster facies Q, further downstream, consisted of increased levels of Fe, Mn, clay, and organic matter which suggested that precipitation processes in the estuary were an important factor. Cluster facies R consisted of polluted sediments.
Mercury levels in various sediments have been presented earlier (table 2.2). The control of the mercury content of estuaries will be discussed later.

Mercury in Estuaries: European Cases. A very high degree of contamination has been observed in the estuaries around the mouths of the Rhine and Maas (De Groot et al, 1973), Tyne, Severn (Bristol channel) (Hamilton et al, 1979), Firth of Forth (Covill, 1971), Wester and Elbe (Forstner & Wittman, 1979), and Humber (Jaffe & Walters, 1977).
As a result of the high input of pollutants from both the British Isles and the continent of Europe, together with the limited dilution due to the shallowness of the water, the southern, compared with the northern, part of the North Sea is one of the most heavily polluted marine areas in the world (Weichart, 1973; Forstner & Wittmann, 1979).
The abundances of 40 major and trace elements in highly polluted sediments from the Rhine river near to the industrial areas around Mainz, West Germany, have been determined by the NAA, and mercury concentrations as high as 94ppm have been reported by Dissanayake et al (1983). Earlier, it was known that the annual load of mercury reaching the Rhine estuary in suspension was 53 tons. The abundance of mercury, together with Cd and Zn, in the Ginsheim sediments exceeds that found in the municipal sewage from American cities (Furr et al, 1976).
In the north-east Irish Sea and Liverpool Bay areas, the levels of mercury in fish have been monitored since 1970 (Portmann, 1979). The annual inputs have been re-estimated by Preston and Portmann (1981) for 1972, from data relating to 1976. Due allowance

was made for major changes in input between 1972 and 1976 and projections have been made accordingly (table 4.6).

Table 4.6: Net Inputs of Mercury to Liverpool Bay
(kg/day - Preston & Portmann, 1981)

Source	1972	1976	1977	1978	1979	1980*
River Mersey input, Howley Weir**	1.0	-	-	-	-	-
direct sewage	0.8	0.8	0.8	0.8	0.8	0.8
dumping (mainly sewage sludge)	3.0	7.4	6.0	2.2	2.2	2.2
dredge spoil (ship canal only)	1.6	0.9	0.9	0.9	0.9	0.6
industrial discharge reaching Mersey estuary	16.0	7.0	5.0	5.0	5.0	3.5
TOTAL	22.4	17.1	13.7	9.9	9.9	8.1

* projected values, ** based upon 1976 data

Using data on the concentration of mercury in fish flesh (MAFF, 1973), an input of 22kg/day was related to a fish flesh mercury level of 0.44mg/kg. Thus, in order to meet an EQS of 0.3mg/kg, the mercury input should not exceed 9kg/day.

The level of mercury in Manchester Ship Canal dredgings is largely due to chlor-alkali plant discharges of about 4.3kg/day. In many industrial areas of the UK, the installation of effluent treatment plants has brought about a considerable reduction in the mercury contamination of estuaries. By 1978, industrial discharges reaching the Mersey estuary directly had been reduced to 5kg/day from 16kg/day in 1972. A further reduction to 3.5kg/day is likely.

The outer Thames estuary data from 1971 have been analysed by Preston and Portmann (1981). The net inputs of mercury to the outer Thames estuary and the actual and predicted mercury contents of fish are shown in tables 4.7 and 4.8, respectively. These tables also show that a mercury input of about 25kg/day was responsible for an increase to a mean mercury concentration, in the flesh of edible fish, of 0.45mg/kg or 0.25mg/kg above the averages for clean and highly contaminated coastal areas, respectively. Thus, in order to observe an Environmental Quality Standard (EQS) of 0.3mg/kg (i.e. 0.1mg/kg above normal), the total input of mercury had to be reduced to approximately 10kg/day.

Table 4.7: Net Input of Mercury to the Outer Thames Estuary
(kg/day - Preston & Portmann, 1981)

Source	1971	1976	1977	1978	1979
sewage effluents	6.8	1.7	1.7	1.7	1.7
river Thames & main tributaries	0.35*	0.35	0.35	0.35	0.35
other water courses	0.20*	0.20	0.20	0.20	0.20
dumping (mainly sewage sludge)	18.0*	3.6	3.8	3.8	4.1
TOTAL	25.4	5.9	6.1	6.1	6.4

* based on 1972 data

In 1971, sewage discharge and sewage dumping accounted for 97% of the input. By introducing sewage treatment plants, the input has been reduced from 25.4kg/day to 5.9kg/day in 1976. This level has been maintained up to the present day. The value is much lower than the EQS value (FAO/WHO, 1974) of 9kg/day, corresponding to 0.3mg/kg of mercury in wet fish flesh.

ROLE OF BIOTA IN MOVEMENT

Microbes play an important role in the movement of mercury in nature, especially in the soil, sediments, and aqueous environments. The main result of microbial action on mercury seems to be its volatilisation, whether it involves reduction of the mercuric ion or methyl- or phenylmercury compounds to volatile Hg^{o}, or whether it involves the conversion of the mercuric ion to dimethylmercury or of the phenylmercuric ion to diphenylmercury. The mercuric ion (Hg^{2+}) may be methylated by bacteria and fungi to give methylmercury [$(CH_3)Hg^+$], which is water soluble. Some bacteria may further methylate methylmercury and convert it to dimethylmercury, which is volatile and escapes into the air.

Table 4.8: Actual and Predicted (in Parentheses) Mercury Content of Fish from the Outer Thames Estuary (mg of Hg per kg of net muscle weight - Preston & Portmann, 1981)

Parameter	1971	1976	1977	1978	1979
		All Fish			
number of fish	385	254	200	130	81
number of samples	14	14	7	7	3
overall mean	0.45	0.30(0.26)	0.21(0.26)	0.81(0.26)	0.81(0.26)
standard deviation	0.2	0.13	0.08	0.05	0.02
range	0.06-2.5	0.05-2.0	0.05-1.1	0.07-1.0	0.05-0.46
		Flounders			
Parameter	1971	1976	1977	1978	1979
number of fish	83	11	100	20	-
mean	0.74	0.46(0.43*)	0.30(0.43*)	0.26(0.43*)	- (0.43*)
range	0.08-2.5	0.10-1.0	0.07-1.1	0.16-0.41	-

* calculated from all fish predicted mean using 1971 flounder/all-fish ratio

Upon weathering, mercuric sulphide (cinnabar, HgS) is converted to mercuric sulphate and becomes disseminated in soil and water. Bacteria, fungi, and humic acid reduce Hg^{2+} to Hg°. Bacteria and fungi may methylate some of the Hg^{2+} and cause a wider range of distribution. Methylmercury, as well as phenylmercury, may again be enzymatically reduced to volatile Hg° by bacteria. This causes detoxification of soil.

Phenylmercury, which is usually of anthropogenic origin, may be reduced by soil bacteria and converted to diphenylmercury. Biogenic H_2S may convert the mercuric ion to HgS, again under anaerobic conditions. The mercury bio-cycle is shown in figure 4.2.

Following the application of mercury fungicides, the metal is transferred to fruits, tubers or seeds in plants (Smart, 1968). Foliar applications of phenylmercuric acetate

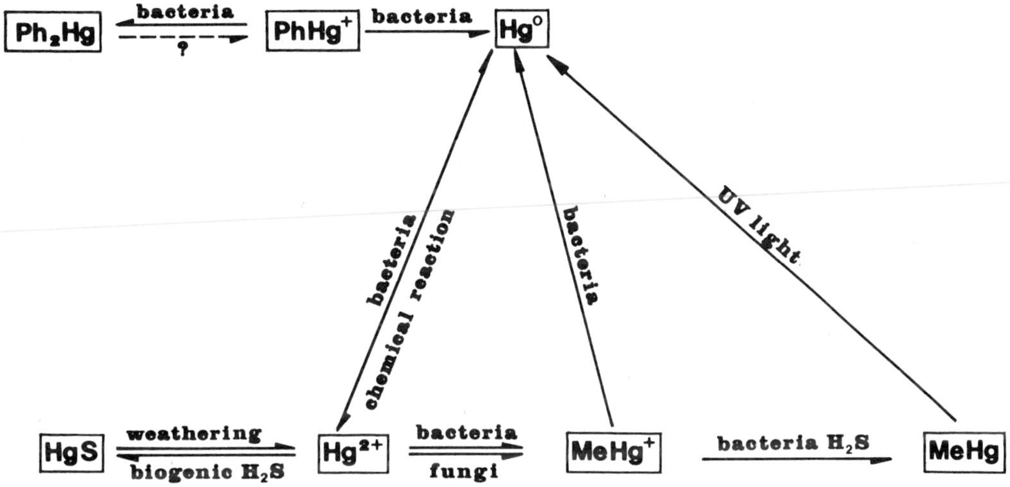

Figure 4.2 Mercury Transformation by
Microbes and Chemical or Physical Agents.

to rice resulted in the transfer of mercury to the grain.

Terrestrial animals accumulate mercury mainly from their food. Birds accumulate mercury mainly by eating seeds and grain treated with mercurial fungicides, etc.

Mercury is transformed into methylmercury, mainly by micro-organisms, and eventually becomes concentrated in large fish (figure 8.1). A very limited conversion of mercury components to methylmercury occurs in hens. A methylmercury-fed hen laid eggs containing only methylmercury (Westöö, 1969). The major portion of mercury present in wild Oregon pheasant tissue was in the form of methylmercury (Buhler et al, 1973) and similar results were observed in Californian pheasants (Griffith, 1973).

Mercury tends to accumulate in the growing feathers, claws, and beaks of birds, and all of the bodily content of mercury is eventually deposited in feathers and other keratinous structures (Study Group on Mercury Hazards, 1970).

Moulting, etc., will release large amounts of mercury. Rapid seasonal changes in the mercury level have been observed in wild pheasants, due possibly to the availability of mercury-treated seeds.

REFERENCES

- Aidin'yan, N.K. (1962) - in Chemical Abstracts, 57, 16336
- Aidin'yan, N.K. (1963) - in Chemical Abstracts, 59, 7237
- Aidin'yan, N.K. and Belavskaya, G.A. (1963) - Chemical Abstracts, 59, 8471
- Aidin'yan, N.K., Troitskii, A.I., and Belavskaya, G.A. (1964) - Geochemistry International, 4, 670-5
- Airey, D. (1982) - Science of the Total Environment, 25, 19-40
- Airey, D. and Jones, P. (1982) - Water Research, 16, 565-77
- Alberts, J.J., Schindler, J.E., Miller, R.W. and Nutter, D.E. (1974) - Science, 184, 895-7
- Anderssen, A. (1967) - Grundförbättring, 20(3-4), 95-105 (English summary)
- Anderssen, A. (1970) - Grundförbättring, 23(5), 31-40 (English summary)
- Anfålt, T., Dyrssen, D., Ivanova, E. and Jagner, D. (1968) - Svensk Kemisk Tidskrift, 80, 340-2
- Appelquist, H.K., Jenson, O., Sevel, T. and Hammer, C. (1978) - Nature, 273, 656-9
- Arnold, M. (1976) - Géochimie du Mercure dans l'Atmosphère et l'Hydrosphère Marine, Report from the Centre des Faibles Radio-activités, Gif-sur-Yvette, France, 69
- Aston, S.R., Bruty, D., Chester, R. and Padgham, R.C. (1973) - Nature, 241, 450-1
- Bogen, J. (1973) - Atmospheric Environment, 7, 1117-25
- Braman, R.S. and Johnson, D.L. (1974) - Environmental Science and Technology, 8, 998-1003
- Brouzes, R.J.P., McLean, R.A.N. and Tomlinson, G.H. (1977) - Research Report, Research Centre of Domtar Ltd., Montreal, Canada
- Brune, D. (1969) - Analytica Chimica Acta, 44, 15-20
- Buat-Menard, P. and Chesselet, R. (1979) - Earth and Planetary Science Letters, 42, 399

- Buhler, D.R., Clayes, R.R. and Rayner, H.J. (1973) - in Mercury in the Western Environment (edited by D.R.Buhler), Proceedings of a Workshop, Portland, Oregon, 25-26th February, 1971, 199-211

- Cambray, R.S., Jeffries, D.F. and Topping, G. (1979) - Marine Science Communications, 5, 175-94

- Carr, R.A. and Wilkniss, P.E. (1973) - Science, 181, 843-4

- Chester, R., Aston, S.R., Stoner, J.H. and Bruty, D. (1974) - Journal de la Recherche Atmosphèrique, 8, 777

- Cole, H.A. (1979) - Proceedings of the Royal Society of London, Series B, 205, 17-30

- Covill, R.W. (1971) - Proceedings of the Royal Society of Edinburgh, 71(B), 143-70

- Crockett, A.B. and Kinnison, R.R. (1977) - US Environmental Protection Agency, Las Vegas, Nevada, EPA-600/3-77-063, May, 1977, 1-8

- Dall'Aglio, M. (1968) - in Origin and Distribution of the Elements (edited by L.H.Ahrens), Pergamon Press, New York, 1065-81

- Dams, R. and De Jonge, J. (1976) - Atmospheric Environment, 10, 1079-84

- Dams, R., Robbins, J.A., Rahn, K.A., and Winchester, J.W. (1970) - Analytical Chemistry, 42, 861

- De Groot, A.J. (1973) - in North Sea Science (edited by E.D.Goldberg), 308-25, MIT Press, Cambridge, Massachusetts

- De Groot, A.J., Allersma, E., and Driel, W.V. (1973) - Symp. Water in Dienst. Van Industrie en Milieu, 110:24

- Dissanayake, C.E., Tobschall, H.J., Palme, H., Rast, U. and Spettel, B. (1983) - Science of the Total Environment, 29, 243-60

- Drake, H.J. (1977) - Mercury, Bureau of Mines, Minerals Yearbook, Volume 1, US Government Printing Office, 189-902

- Dumarey, R., Heindryckx, R., Dams, R., and Hoste, J. (1979) - Analytica Chimica Acta, 107, 159-67

- Eriksson, E. (1967) - The Mercury Problem., Oikos, Supplement 9, 13-20

- FAO (1978) - Yearbook of Fishery Statistics, Volume 44, FAO-UN, Rome, 342pp

- FAO/WHO (1974) - WHO Technical Report Series, No. 555

- Ferm, R. and Larsson, J.E. (1973) - Kricksilver-anvandring, Kontroll och Miljoeffekter Statens Naturvardsvesk, PM421, 8-17

- Fitzgerald, W.F. and Gill, G.A. (1979) - Analytical Chemistry, 51, 1714

- Fleischer, M. (1970) - in Mercury in the Environment, Geological Survey Professional Paper 713, 6-13

- Fogg, T.R. and Fitzgerald, W.F. (1979) - Journal of Geophysical Research, 84, 6987-9

- Forbes, E.A., Posner, A.M. and Quirk, J.P. (1974) - Journal of Colloid and Interface Science, 49, 403-9

- Forstner, U. (1977) - in Interactions between Sediments and Fresh Water, Wageningen, The Hague, Pudoc/Junk B.V., 94-103

- Forstner, U. and Wittman, G.T.W. (1979) - Metal Pollution in the Aquatic Environment, Springer Verlag, 475

- Frenet-Robin, M. and Ottmann, F. (1978) - Estuarine and Coastal Marine Science, 7, 425-36

- Friberg, L. and Vostal, J. (1972) - in Mercury in the Environment (edited by L.Friberg & J.Vostal), Chemical Rubber Company Press, Ohio, 183-6

- Frissel, M.J., Poelstra, P. Reiniger, P. and Das, H.A. (1971) - in Radioecology Applied to the Protection of Man and his Environment, Proceedings of an International Symposium, Rome, 941-55

- Frykholm, K.O. (1970) - Svensk Tandlazkare Tidskrift, 63, 763-72

- Furr, A.K., Lawrence, A.W., Tong, S.S.C., Grandolfo, M.C., Hofstader, R.A., Bache, C.A., Gutenmann, W.H. and Lisk, D.J. (1976) - Environmental Science and Technology, 10, 683-7

- Furzov, V.Z. (1970) - American Geological Institute, Doklady Academy of Sciences USSR, 194, 209-11

- Gardner, D. (1975) - Marine Pollution Bulletin, 6, 43-6

- Gardner, D. and Riley, J.P. (1973) - Nature, 241, 526-7

- Ginzburg, I.I. (1960) - Principles of Geochemical Prospecting, Pergamon Press, London (translated from Russian)

- Goldberg, E.D. (editor) (1976) - The Health of the Oceans, Unesco Press

- Great Britain (1971) - Parliamentary Working Party on the Monitoring of Foodstuffs for Mercury and other Heavy Metals, First Report Survey of Mercury in Food, London, HMSO

- Griffith, W.H. (1973) - in Mercury in the Western Environment (edited by D.R.Buhler), Proceedings of a Workshop, Portland, Oregon, 25-26th February, 1971, 135-9
- Haines, S.K. (1977) - Barite, in Minerals Year Book, 1975, Volume 1, Bureau of Mines, US Government Printing Office, Washington, 1977, 225-31
- Hamaguchi, H., Rokuru, K., and Hosohara, K. (1961) - Nippon Kagaku Zasshi, 82, 347-9, Chemical Abstracts 55, No.1522C
- Hamilton, E.I., Watson, P.G., Cleary, J.J., and Cliffon, R.J. (1979) - Marine Geology, 31, 139-82
- Hawkes, H.E. and Williston, S.H. (1962) - Mining Congress, 48(12), 30-2
- Heide, F., Lerz, W., and Boehm, G. (1957) - Naturwissenschaften, 44, 441-2
- Hem, J.D. (1970) - in Mercury in the Environment, USGS Professional Paper 713, 19-24
- Hogstrom, U., Enger, L., and Svedung, I. (1979) - Atmospheric Environment, 13, 465-76
- Hosohara, K. (1961) - Nippon Kagaku Zasshi, 82, 1107-8; Chemical Abstracts (1962), 56, 4535
- Hosohara, K., Kozuma, H., Kawasaki, K., and Tsurata, T. (1961) - Nippon Kagaku Zasshi, 82, 1479-80, Chemical Abstracts (1962), 56, 57669
- Hughes, H.L. (1957) - Annals of the New York Academy of Sciences, 65, 454-60
- Jaffe, D. and Walters, J.K. (1977) - Science of the Total Environment, 7, 1-15
- James, C.H. (1962) - A Review of the Geochemistry of Mercury and its Application to Geochemical Prospecting, Imperial College of Science and Technology, Technical Communication No. 41, (unpublished), 42
- Jenson, S. and Jernelov, A. (1969) - Nature, 223, 753-4
- Jepson, A.F. (1973) - Advances in Chemistry, 123, 81-95
- Jernelöv, A. and Wallin, T. (1973) - Atmospheric Environment, 7, 209-14
- Joensüü, O.I. (1971) - Science, 172, 1027-8
- Johansson, I.R. (1970) - Mercury in the Natural Environment: a Review of Work, Geological Survey of Canada Paper, 70-7
- Johansson, I.R. and Boyle, R.W. (1971) - in Proceedings Soc. Canada, Symposium on Mercury in Man's Environment, Ottawa, 5-21

- Johnson, D.L. and Braman, R.S. (1974) - Environmental Science and Technology, 8, 1003-9

- Karasik, M.A. and Bol'shakov, A.P. (1965) - Doklady Academii Nauk USSR, Earth Science Section (English translation), 161, 204-6

- Khayretdinov, I.A. (1971) - Geokhimia, 6, 668-83

- Klein, D.H. and Russel, P. (1973) - Environmental Science and Technology, 7, 357-8

- Kokshoy, M. and Bradshaw, P.M.D. (1969) - Colorado School of Mines Quarterly, 64(1), 333-56

- Kothny, E.L. (1973) - Trace Elements in the Environment (edited by E.L.Kothny), Advances in Chemistry Series, 123, 48-81

- Kozuchowski, J. and Johnson, D.L. (1978) - Nature, 274, 468-9

- Lantzy, R.J. and MacKenzie, F.T. (1979) - Geochimica et Cosmochimica Acta, 43, 511-25

- Luoma, S.N. (1983) - Science of the Total Environment, 28, 1-22

- MacKenzie, F.T. and Wollast, R. (1977) in - The Sea (edited by E.D.Goldberg, J.N.McCave, J.J.O'Brien, & J.H.Steele), volume 6, Marine Modelling, John Wiley and Sons, New York, 765-77

- MAFF (1973) - Working Party on the Monitoring of Foodstuffs for Heavy Metals, Third Report Survey of Mercury in Food: a Supplementary Report, HMSO, London

- Matsunaga, R. and Goto, T. (1976) - Geochemistry Journal, 10, 107-9

- McCarthy, J.H. (1972) - Journal of Geochemical Exploration, 1, 143-62

- McCarthy, J.H., Meuschke, J.L., Flickin, W.H., and Learned R.E. (1970) - Geological Survey Professional Paper 713, US Government Printing Office, 37-9

- McCarthy, J.H., Vaughn, W.W. Learned, R.E., and Meuschke, J.L. (1969) - US Geological Survey Circular 609, US Department of the Interior, Washington, 16

- McNerney, J.J., Buseck, P.R., and Hauson, R.C. (1972) - Science, 178, 611-2

- McNerney, J.J. and Buseck P.R. (1973) - Economic Geology, 68, 1313-20

- Metallgesellschaft AG (1978) - Metal Statistics, 1967-1977, A.M.Main, Frankfurt, 65th Edition

- Meyer, W.T. and Evans D.S. (1973) - in Prospecting in Areas of Glaciated Terrain (edited by M.J.Jones), 127-38

- Millard, G.E. and Griffin, J.H. (1980) - Science of the Total Environment, 16, 239-43
- Miller, D.R. and Buchanan, J. (1979) - in A Technical Report, MARC, No.14, Chelsea, London
- Millward, G.E. and Herbert, I. (1981) - Environmental Pollution (Series B), 2, 265-74
- National Academy of Sciences (1977) - An Assessment of Mercury in the Environment, National Academy of Sciences, Washington, D.C.
- Neville, G.A. (1967) - Canadian Chemical Education, 3(1), 4-7
- Nishimura, M. (1979) - in ACS/CSJ Chemical Congress, INOR, 175 Honolulu, Hawaii
- Phillips, G.P., Dixon, B.E., and Lidzey, R.G. (1959) - Journal of the Science of Food and Agriculture, 10, 604-10
- Piotrowski, J.K. and Coleman, D.O. (1980) - Environmental Hazards of Heavy Metals: Summary Evaluation of Lead, Cadmium and Mercury, MARC Report No. 20
- Porritt, R.E. and Swaine, D.J. (1976) - preprint Inst. Fuel Conf. Sydney, Paper 18,
- Portmann, J.E. (1979) - Aquatic Environment Monitoring Report, MAFF Direct. Fish Res. Lowestoft, No.1
- Preston, A. and Portmann, J.E. (1981) - Environmental Pollution (Series B), 2, 451-64
- Rahn, K. (1976) - University of Rhode Island Technical Report, 265
- Rahn, K. (1980) - Conference on the Arctic Ocean, London, 11-12th March
- Ramamoorty, S. and Rust, B.R. (1978) - Environmental Geology, 2, 765-72
- Rogers, R.D. (1977) - Journal of Environmental Quality, 6, 463-7
- Ross, R.G. and Stewart, D.K.R. (1962) - Canadian Journal of Plant Science, 42, 280-91
- Sadasivan, S. (1981) - Science of the Total Environment, 20, 109-15
- Saukov, A.A. (1946) - Trudu Instituta Geologicaeskogo, Akademii Nauk SSR, 78(17), 129
- Saukov, A.A. (1953) - Geochemie, VEB Verlag Technik, Berlin, 76

- Schlesinger, W.H., Reiners, W.A., and Knapman, D.S. (1974) - Environmental Pollution, 6, 39-47

- Sergeev, E. (1961) - International Geological Review, 3, 93-9

- Sevel, T. (1975) - in Seminar on Tungmetaller, Sirkulasjon i jordbruget, 21-23rd October, As-Trykk, Oslo, 84-93

- Sidgwick, N.V. (1950) - The Chemical Elements and their Compounds, Oxford University Press, London, Volume 1, 324

- Siegel, S.M., Siegel, B.Z., Puerner, N., and Speitel, T. (1975) - Water, Air, Soil Pollution, 4, 9-18

- Slemr, F., Seiler, W., Eberling, C., and Roggendorf, P. (1979) - Analytica Chimica Acta, 110, 35-47

- Smart, N.A. (1968) - Residue Review, 23, 1

- Stahl, O.R. (1969) - Preliminary Air Pollution Survey of Mercury and its Compounds: a Literature Review, US Department of Health and Welfare, Public Health Service, National Air Pollution Control Administration

- Stevens, D.N., Bloom, D.N., and Bisque, R.E. (1969) - Colorado School of Mines Quarterly, 64(1), 513-4

- Stock, A. and Cucuel, F. (1934) - Naturwissenschaften, 22, 390-3

- Straby, A. (1968) - appendix in Report to the Swedish Technical Research Council, Isotope Technics Laboratories, S-114, 28 Stockholm, Stencils

- Study Group on Mercury Hazards (1971) - Environmental Research, 4, 1-69

- Symon, D. (1962) in - Equilibrium Diagrams for Minerals at Low Temperatures and Pressures (edited by H.H.Schmitt), Geological of Harvard, Cambridge, 164-75

- Trost, P.B. and Bisque, R.E. (1972) - in Environmental Mercury Contamination, (edited by R.Hartung & B.D.Dinman), Ann Arbor Scientific Publications Incorporated, Ann Arbor, Michigan, 178-96

- Vaughn, W.W. (1967) - US Geological Survey Circular, 540, 1-8, US Government Printing Office, Washington DC

- Warren, H.V. and Delavault, R.E. (1969) - Oikos, 20, 537-9

- Weichart, G. (1973) - Naturwissenschaften, 60, 469-72

- Weiss H., Bertine, K., Koide, V.M., and Goldberg, E.D. (1975) - Geochimica et Cosmochimica Acta, 39, 1-10

- Weiss, H., Koide, V.M., and Goldberg, E.D. (1971) - Science, 174, 692-4

- Westermark, T. and Ljungreen, K. (1968) - in Report to the Swedish Technical Research Council, Isotope Technics Laboratories, S-114 28 Stockholm, Stencils

- Westöö, G. (1967) - Acta Chimica Scandinavica, 21, 1790-800

- Westöö, G. (1969) - in Chemical Fallout, (edited by M.W.Miller & G.G.Berg), Charles Thomas Springfield, Illinois, p.75

- Wicklander, L. (1969) - Geoderma, 3, 75-9

- Williston, S.H. (1964) - Engineering and Mining Journal, 165(5), 98-101

- Williston, S.H. (1968) - Journal of Geophysical Research, 73, 7051-5

- Windom, H.L. (1973) - Proceedings of American Society of Civil Engineering, 99(WW2), 257

- Windom, H.L. and Taylor, F.E. (1979) - Deep Sea Research, 26(3A), 283-92

- Windom, H.L., Taylor, F.E., and Waiters, E.M. (1975) - Deep Sea Research, 22, 629-33

- Yamada, M. and Tonomura, K. (1972) - Journal of Fermentation Technology, 50, 159-66

- Zautashvili, B.Z. (1966) - Geokhimiya, 357-62 (in Russian); Chemical Abstracts, 64, 17267

CHAPTER V

ANALYSIS OF MERCURY

INTRODUCTION

Interest in geochemical-cum-environmental studies, coupled with the realisation that metallic mercury can be converted into dimethylmercury and methylmercury compounds and appear in the food chain, has led to the accelerated development of rapid and accurate qualitative methods for the analysis of mercury in organic and inorganic substances. The sampling and analysis of mercury in environmental and biological samples pose some extremely challenging problems. The minute quantitites of mercury present in the sample, as well as the volatile nature of mercury and its associated compounds, add to the problems associated with the complexity of the matrices themselves. Numerous methods for the analysis of mercury are to be found in the literature. A large number of workers have described the methods employed in mercury determination e.g., atomic absorption (Barringer, 1966; Ward, 1970; Ward et al, 1969; Jonasson et al, 1973), cold vapour atomic absorption (Olafsson, 1974; Stoeppler et al, 1979), neutron activation analysis (Pillay et al, 1971; Zmijewska, 1977; Litman et al, 1977; Jensen & Carlson, 1978), gas-liquid chromatography and, to a lesser extent, gravimetry, titrimetry, thin-layer chromatography, X-ray fluorescence, pyrolysis, isotope exchange, gold microbalance, gold resistivity (McNerney & Buseck, 1973), and differential pulse anodic stripping voltammetry (DPASV) (Andrews et al, 1976; Sipos et al, 1979; Ahmed et al, 1981). Methods of mercury analysis have been reviewed by Ure and Shand (1974).

Many workers estimate the total mercury content by neutron activation or use flameless atomic absorption with neutron activation, and/or gas chromatography in order to determine inorganic or organic mercury species in biological materials, with or without isotopic labelling (Becknell et al, 1971; Dolar et al, 1971; Fang, 1973; Jones & Nickless, 1974; Windom et al, 1976). Isotopic labelling has been used to provide a quantitative determination of the mercury in biological materials (Clarkson & Greenwood, 1970; Becknell et al, 1971; De Filippis, 1978). The use of anion exchange filters (Becknell et al, 1971), chelating resins, mercury traps, and some forms of thin-layer chromatography (TLC) have been suggested for the separation of mercury species (Gay, 1976). Tissue extraction efficiencies range from 5% (Fang, 1973) to 95% (Clarkson & Greenwood, 1970) in animal tissues. Animal or plant tissues which have been digested at temperatures of up to 85°C give recoveries of up to 99-100% (Ben-Bassat & Mayer, 1977). This level of recovery is easily attained in water samples (Becknell et al, 1971). A modified technique has been employed at the National Research Council of Canada for the more effective analysis of sediments, fish, and water (Kudo et al, 1977a,b; Akagi et al, 1979).

Normal techniques of optical spectroscopy and spectrophotometry are not sufficiently sensitive to determine mercury in silicate rocks. Neutron activation analysis gives good results (Ehman & Lovering, 1967; Kasper & Kral, 1958). For rocks containing high concentrations of mercury, Popea and Jemaneanu (1960) have proposed a method based upon dithizone extraction into carbon tetrachloride solution. Here, 1 to 5g of the sample are refluxed with sulphuric and nitric acids and the resultant mercury dithizonate is extracted at pH 4-5 from acetic acid solution containing both EDTA and potassium thiocyanate. This method is useful for recovering 2 to 20 μg of mercury.

A more detailed discussion of the analysis of geological materials is presented later in this chapter.

The mercury present in biological tissues is mostly in the form of organic mercurials and therefore requires the degradation of the organic materials and/or the careful extraction of mercury from the insoluble matrices. Because of the high volatility of mercury compounds, ordinary combustion processes are not suitable for decomposing the bound mercury in order to collect it for analysis. Wet oxidation processes are often used to degrade the materials for trace analysis.

The use of radioactive tracer techniques in determining recovery and analytical accuracy is not well-suited to the analysis of mercury in biological and environmental studies because of the limited knowledge of, and possible variations in, the chemical form of the mercury present in the samples.

The rate of absorption of methylmercury has been studied in fresh water fish by using tracer methods (Stary et al, 1980; Stary & Kratzer, 1980)

A. COLORIMETRIC METHODS

Colorimetric methods have been used since classical times. These methods have the advantage of being simple, cheap, and less time-consuming than other methods. No special equipment other than a spectrophotometer is needed.

1. DITHIZONATE METHOD

Colorimetric methods using dithizone have gained wide acceptance and are detailed below according to the technique followed by the Association of Agricultural Chemists (1965). This method is suitable for the determination of mercury in fruits, vegetables, beverages, cereals, grain, fish, meat, etc., and can also be employed for other types of sample by modifying the sample treatment.

Dithizone (diphenyl thiocarbazone) reacts with mercury to form four coloured complexes. For analytical use, the primary compound with mercury (II):

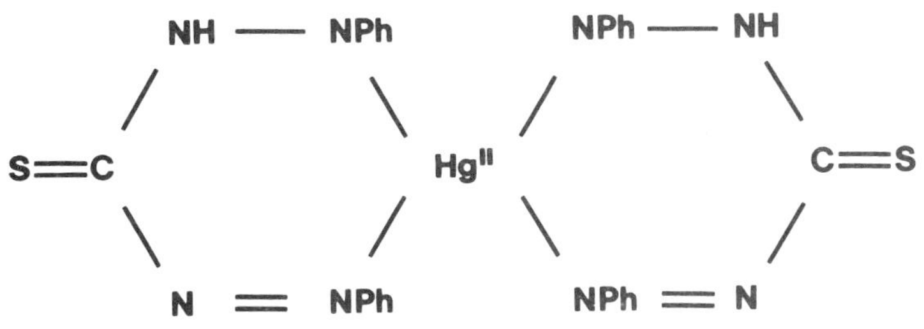

is usually formed by shaking the test solution with a small excess of a carbon tetrachloride- or chloroform-based solution of the reagent. In these solvents, the solution is orange/yellow, with a maximum absorbance at 490nm. Methylmercuric dithizonate has its maximum absorbance at 476nm in chloroform.

The procedures for sample preparation and analysis are detailed below:

(a) Sample preparation; fresh fruits, vegetables, and beverages:

A weighed amount of the sample is digested with nitric acid under a controlled flame so that the digest just refluxes. Refluxing is continued until the digest does not change its consistency. After cooling, a mixture (1:1) of nitric acid and sulphuric acid is added and heating is continued until all of the fibrous material is digested. The digest is then cooled and chilled under cold water so that the fats and waxes solidify. The filtrate is separated and again digested under a small flame with small additions of nitric acid until the digest is colourless or pale yellow.

(b) Dried fruits, cereals, seeds, and grain:

In these cases, the samples should be taken into water before adding nitric acid and proceeding as in (a).

(c) Meats, fish, and biological materials:

These materials contain a large amount of fat and protein. Therefore, the digestion must be conducted carefully in order to avoid foaming. Otherwise, the procedure is the same as in (a).

The sample solutions thus prepared should be titrated against standard alkali, and the acidity should be reduced to 1 N by the addition of a calculated quantity of NH_4OH.

Isolation of Mercury:

The sample solution is transferred to a separator and shaken with dithizone solution. The layers are allowed to separate and the chloroform layer is then taken into another separator containing 0.1 N HCl. This extraction is repeated until the chloroform contains an excess of dithizone.

The extract is then shaken vigorously for 1 min with sodium thiosulphite solution and the chloroform layer is completely drained off and discarded. It is then again shaken vigorously with sodium hypochlorite in order to decompose the mercury thiosulphate complex and to oxidise excess thiosulphate.

Determination of mercury:

To the resulting solution, 3ml of dilute acetic acid and an appropriate volume of dithizone solution are added and shaken

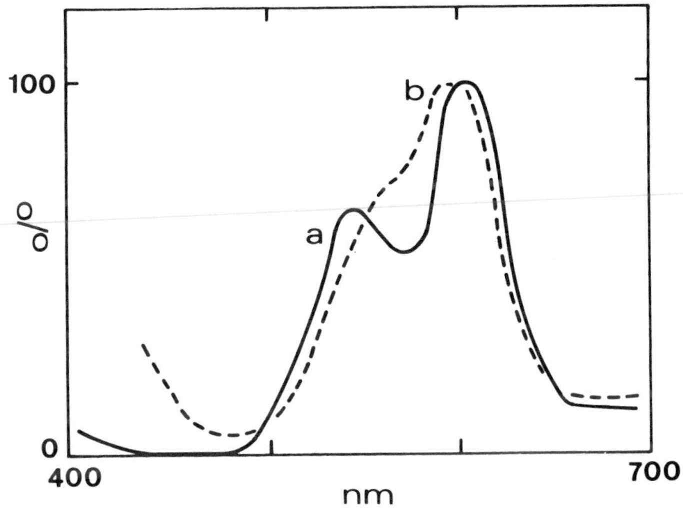

Figure 5.1 Absorption Spectra of Crystal Violet Complexes. The Relative absorption is expressed as a percentage of the peak value at a wavelength of 605nm (a: mercury complexes, b: mercury interference due to incomplete oxidation of the sample)

for 1 min. The layers are allowed to separate and the dithizone extract is collected, filtered through cotton pledgets, and transferred to photometric cells. The absorbance at 490nm is determined and the mercury concentration is deduced by comparison with a standard curve.

Kamath et al (1975) developed a rapid method for the determination of mercury in sediments using the dithizone method. The sediment sample was dried and intimately mixed with ammonium chloride. The mixture was gently heated so that the mercury formed volatile compounds. The vapours were absorbed in sulphuric acid and the mercury was separated as a dithizone complex and determined spectrophotometrically.

For water samples, the following simple procedure may be sufficient:

100ml of the sample are acidified to a pH of between 0 and 1, using dilute H_2SO_4, and extracted twice with 10ml of dithizone solution. If the extract does not have a green colour, smaller samples are taken. The combined extracts are washed three times with ammonia solution (5ml in 1 litre of water), and any ammonia remaining is neutralised with dilute acetic acid. The extract is then filtered into a 1cm spectrophotometry cell and the absorbance of the orange complex at 485nm is measured. The calibration range is 5 to 30 μg/g of mercury.

2. CRYSTAL VIOLET METHOD

This is a solvent extraction method (Kothny, 1970) for mercury determination which uses crystal violet as the colorimetric reagent. The absorption spectra are shown in figure 5.1.

Principle:

The mercury is directly complexed with iodine. The excess oxidant is destroyed with sulphite and the complex mercury tetraiodide ion is converted to the disulphitomercurate diiodide, which is resistant to reducing and oxidising compounds. Then, EDTA is added in order to complex interfering metals. Ethyleneglycol monoethyl ether is added to desorb and coagulate colloids which may form during complexation or extraction. The solution is made to contain 0.01% of crystal violet (Cv), upon which the complex, $(Cv)HgI_3$, precipitates. The compound is dissolved in toluene and the absorbances at the absorption maxima near to 605nm are measured by spectrophotometry.

3. THIOSALICYCLAMIDE METHOD

Mazumdar and Shome (1971) used thiosalicylamide for the determination of mercury:

An aliquot of mercury (II) solution was taken in a 25ml flask and the pH was adjusted to 2.5 using sodium acetate and hydrochloric acid. Then, 2.0 to 3.5ml of 0.02M thiosalicylamide solution was added, thoroughly mixed, and the volume was made up with water and ethanol so that the final solution contained 50% ethanol. The absorbance at 355nm was determined, after 30 min, against a reagent blank.

The mercury thiosalicylamide complex is stable and can be obtained in a directly weighable form. Therefore, with the help of thiosalicylamide, mercury can also be determined gravimetrically.

4. BRILLIANT (C.I. BASIC) GREEN METHOD

Sawaya et al (1973) have suggested a method in which the ion-association complex of Brilliant Green with HgI_4^{2-} can be extracted from an aqueous medium (pH: 0.7 to 1.5) into benzene. It contains mercury and reagents in the ratio, 1:2, and the absorption maximum is at 640nm. Perchlorate, SCN^-, and Fe (III) seriously interfere with the determination of mercury using this method.

Procedure:

To 21ml of solution, containing up to 20 μg of Hg (II) and H_2O, are added 3ml of 0.1M KI and 3ml of Brilliant Green. The pH is adjusted to between 0.7 and 1.5 using 0.1M H_2SO_4, and the solution is diluted to 30ml and extracted with 10ml of benzene. The extract is dried over Na_2SO_4 and the extinction at 640nm is measured and compared with a standard curve.

5. BRILLIANT (C.I. DIRECT) YELLOW METHOD

Iarayan et al (1973) have suggested a method, using brilliant yellow, for the determination of mercury.

Procedure:

To 10ml of test solution containing 0.1 to 3.4 μg of Hg (II) per ml and 0.01M KBr, 0.2mM brilliant yellow and 15mM to 0.75M H_2SO_4, are added 10ml of extractant. The mixture is shaken for 1 min and the extinction of the extract at 640nm is measured. There is no interference in the presence of a substantial amount of copper or zinc.

Dinaphthyl carbazone forms a mercuric compound which is similar to the dithizonate

(Supronovich, 1936), and has been used for the analysis of biological material (Cholak & Hubbard, 1946) and urine (Hubbard, 1940). 't is discussed in detail by Thabet and Tabibian (1966).

Colorimetric methods are not suitable for the determination of inorganic mercury and methylmercury separately. Moreover, techniques requiring the digestion of samples do not permit inorganic mercury and organic mercury to be distinguished. Thus, before digestion, organomercurials have to be extracted with benzene and determined either by titrimetric methods or, better, characterised by chromatographic methods such as TLC or GLC.

B. TITRIMETRIC METHODS

A method of determining phenyl- and methylmercury salts in biological materials has been developed by Gage (1961). Here, the substance is extracted by benzene from a homogenate of the sample which has been made strongly acidic using hydrochloric acid. The benzene solution is shaken with a dilute aqueous solution of sodium sulphide, with which the organic mercurials react to form water-soluble complexes. The aqueous layer is then oxidised with acidic permanganate, and the mercury content is determined using the titrimetric method.

Extraction of organic mercury:

> To 20ml of a 5% water homogenate of tissue were added 4ml of HCl and 20ml of benzene. This was shaken for 5 min, and the layers were allowed to separate. The upper benzene layer was transferred and shaken with 10ml of sodium sulphide solution (10% w/v). The layers were allowed to separate and the lower layer was taken.

Oxidation with acid permanganate:

> To the aqueous sulphide solution were added 2ml of 50% v/v H_2SO_4, 0.5g of $KMnO_4$, and 1 drop of octanol. The tube was placed in a boiling bath of water for 1h. It was allowed to cool a little and 3ml of hydroxylamine hydrochloride solution (20%) were added for decolorisation.

Determination:

To the oxidised solution were added 2ml of 10% urea solution and 1ml of 2.5%w/v EDTA solution. The pH was adjusted to 1.5 using sodium acetate, and 0.5ml of chloroform were added. A control solution was prepared by mixing 15ml of water, 2ml of 50%v/v H_2SO_4, 0.5g of $KMnO_4$, and 3ml of $NH_2OH.HCl$.

Dithizone solution (a 20 fold chloroform dilution of a stock solution of 20mg/100ml of CCl_4) was added slowly to the test solution until a greyish colour was obtained (intermediate between the orange of the mercury complex and the green of the unchanged solution). The volume of dithizone solution used was recorded and the same volume was added to the control solution. The control solution was titrated with standard mercury solution until the colour matched that of the test solution.

C. TLC and GLC

Thin-layer chromatography and gas-liquid chromatography are now widely used for the determination of mercury, and are particularly suitable for the analysis of organomercurials. Solvent extraction clean-up makes use of the ease with which the anions of water-soluble salts can be exchanged (Rodriguez-Vazquez, 1978).

Westöö (1966, 1967) used a combined (gas chromatographic and thin-layer chromatographic) method for the determination of methylmercury compounds in fish, eggs, meat, and liver as outlined below.

Extraction and Purification:

Tissues were homogenised with water and acidified with concentrated HCl. Organomercurials were extracted in one step with benzene. From an aliquot of the benzene solution, organomercury could be extracted with ammonia or sodium hydroxide solution, saturated with Na_2SO_4 for the elimination of lipids, and re-extracted with benzene after acidification with HCl. The methylmercury chloride was transformed to methylmercury dithizonate by concentrating the benzene solution under reduced pressure and adding 0.4%

dithizonate solution in benzene until a green coloration was obtained. Methylmercury dithizonate and phenylmercury dithizonate could be separated from each other by thin-layer chromatography on aluminium oxide (0.2 μg detection limit). The plates were activated at 105°C for 1h, and the developing solvent was a 70/30 mixture of light petroleum and diethyl ether. Gas-liquid chromatography was carried out using a 5ft by 0.125in stainless steel column containing carbowax 20M, a gas flow rate of 65ml N_2/min, a column temperature of between 138 and 145°C, and an injector temperature of between 150 and 170°C.

In the above procedure, the extraction with alkali was incomplete unless the benzene extract had been previously concentrated by distillation. The distillation procedure was assumed to remove volatile thio-compounds which bound part of the methylmercury and prevented its uptake into ammonia. Thus, any methylmercury attached to a sulphur atom of a non-volatile compound, which gave rise to alkali-insoluble methylmercury salts at the purification stage, would not be determined.

Two alterations to the method have been proposed: (a) the mercuric chloride modification, and (b) the cysteine modification. The latter is relatively simple and is described below.

A definite weight of the sample was homogenised with water and transferred to a separation funnel. Concentrated HCl was added to the homogenate, followed by benzene. The mixture was shaken vigorously for 5min. and centrifuged. About 50ml of the extract were transferred to another separation funnel. Here, 6ml of a 1% solution of cysteine acetate saturated with sodium sulphate were added and shaken vigorously. Then, 5ml of the aqueous phase were separated off and acidified with 3ml of 6N HCl, and the solution was extracted with benzene. The extract was dried with anhydrous Na_2SO_4 and submitted to gas chromatography and, after concentration, to thin-layer chromatography. The latter method, but involving dithizonate, has also been used by Pawlaczyk and Sierzant (1974) for the determination of several metals (Hg^{2+}, Zn^{2+}, Bi^{3+}, Ni^{2+}, Co^{2+}, Cd^{2+}) as contaminants of pharmaceutical preparations.

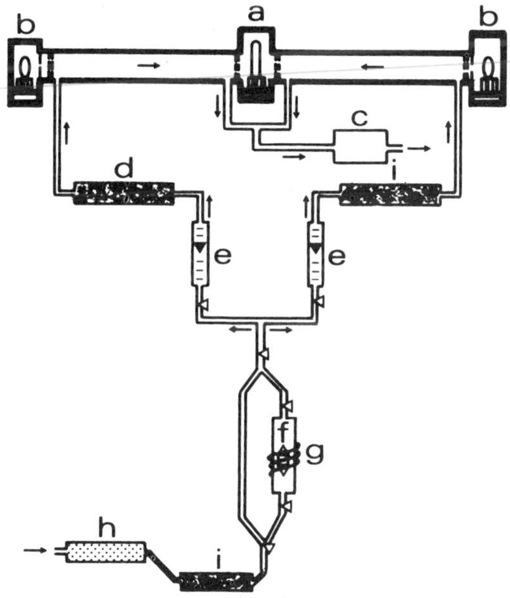

Figure 5.2.1 Resonance Absorption (253.65nm) Mercury Analyser. (a: UV-lamp, b: photocell, c: suction pump, d: glass wool, e: flow meter, f: sample, g: induction heating coil, h: silica gel, i: glass wool with $PdCl_2$)

Several workers have determined organomercurials according to the method of Westöö, using different alkylating or arylating reagents, extracting the organomercurials into benzene, and determining their content by means of GLC. Paul et al (1973) have obtained the best results using Li pentafluorobenzene sulphinate as an arylating reagent, while Zarnegar and Mushak (1974) have used pentacyalus (methyl) cobaltate^{III} and triphenylborate.

For the determination of methylmercury in aqueous solution, a convenient procedure to use is as follows:

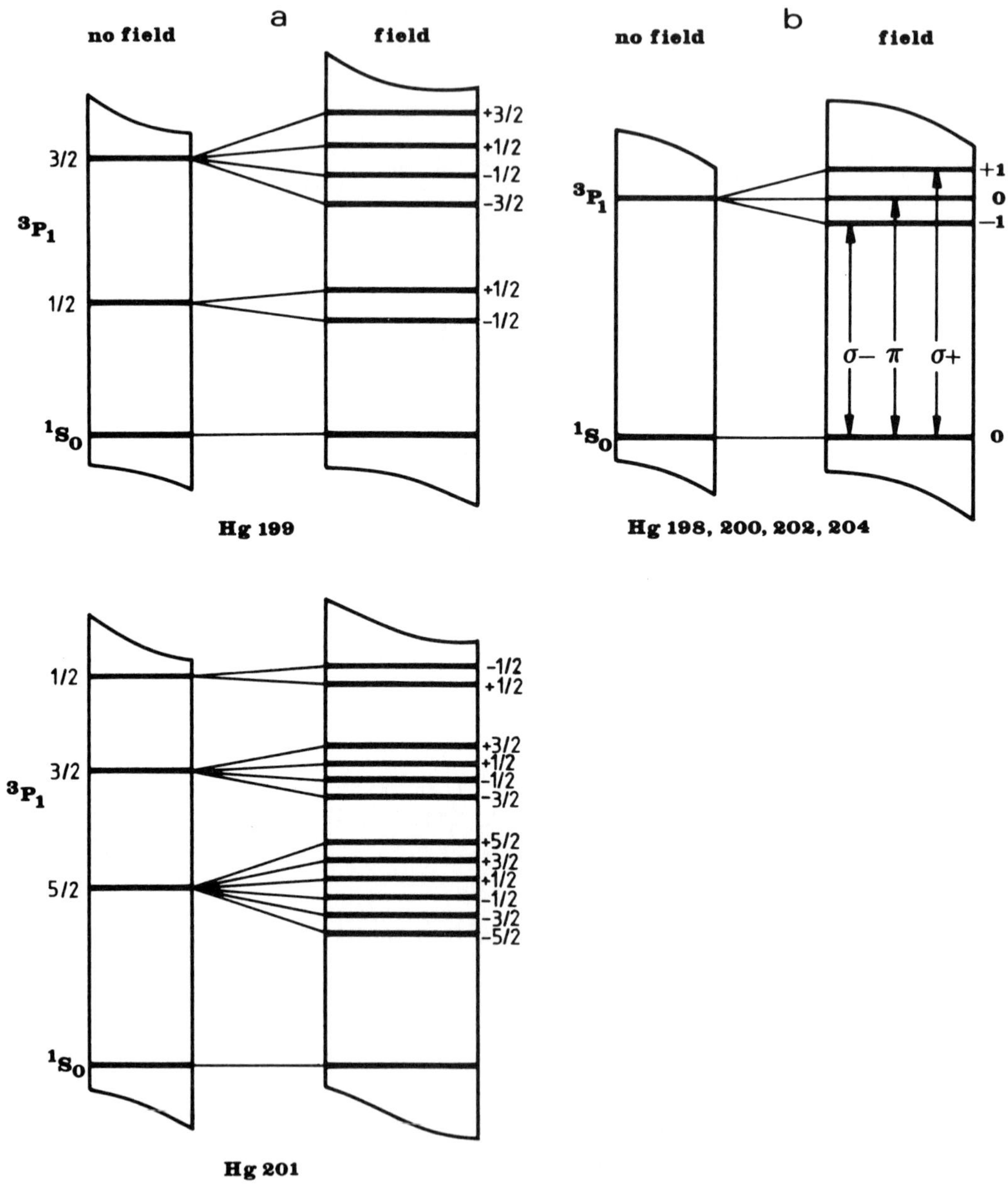

Figure 5.2.2 Zeeman Sub-Levels of Natural Mercury

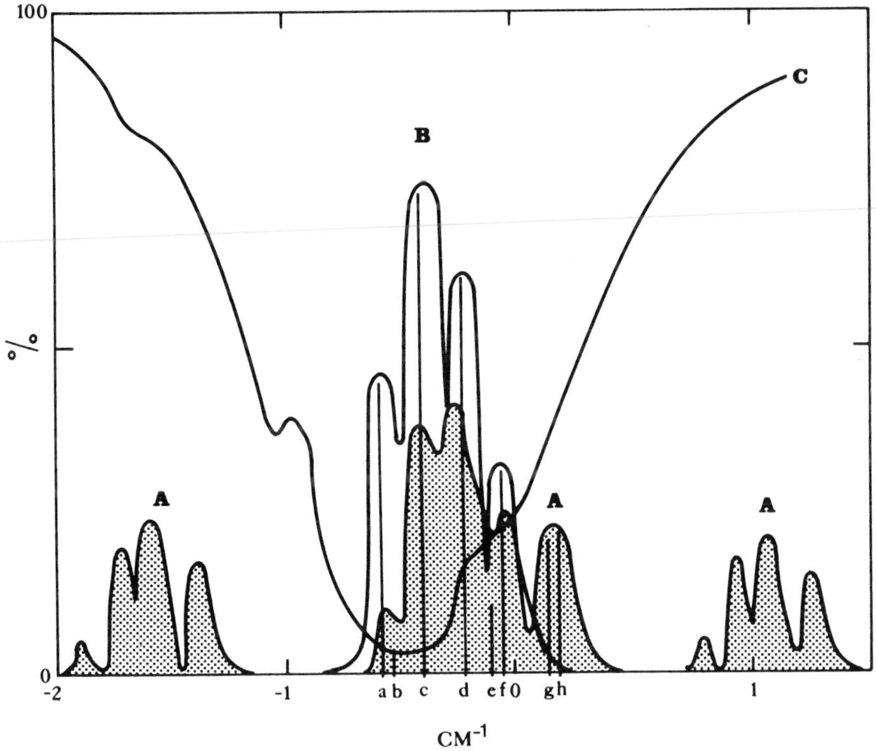

Figure 5.2.3 Emission and Absorption Profiles of Natural Mercury. [Transmission of Hg cell at 1 atmosphere. A: emission profile with field, B: emission profile without field, C: absorption profile, a: 204, 199 (1/2), b: 201 (5/2), c: 202, d: 200, e: 201 (3/2), f: 198, g: 199 (1/2), h: 201 (1/2)]

The sample is treated with KBr and the CH_3HgBr is extracted into toluene. The extract is separated and treated with aqueous thiosulphate in order to strip methylmercury back as $CH_3HgS_2O_3^-$. The addition of iodide gives CH_3HgI, which is volatile and stable enough to be determined by GLC on a glass polar column followed by electron-capture detection.

About 0.1 μg/ml of Hg in benzene extract can be determined using this procedure.

Gas chromatography and thin layer chromatography are fairly accurate methods and have the advantage that both inorganic and organic mercury can be determined sequentially by extracting and analysing two aliquots of a sample.

D. AAS

This technique is the most common choice for the analysis of mercury when neutron activation and other facilities are not available. In this method, the absorption (by ground-state mercury atoms liberated from the sample) of the radiation from a mercury (or hydrogen) vapour discharge (having a predominant 253.652nm line) is measured. This is done by reducing the oxidised mercury samples to the metallic state and transporting the vapour through the optical beam in the absorption cell of the spectrometer (figure 5.2.1; Mitra, 1970).

There are two possible lines available for absorption. The 184.96nm line has an oscillator strength which is about 50 times greater than the other line at 253.65nm. However, the former line is much more difficult to use because the atmosphere and coal gases absorb strongly at this short wavelength.

In common spectrometers, the 253.65nm resonance line is used for Hg detection. A dual-beam dual-wavelength spectrometer has been developed which uses the 253.65nm resonance line. The dual or reference wavelength is generated by Zeeman-splitting* the emission line by repetitively pulsing a magnetic field around a discharge lamp.

---

* A magnetic field brings about splitting of the energy levels involved in a particular transition (figure 5.2.2a). The $^3P_1$-^1S_o transitions for ^{198}Hg, ^{200}Hg, ^{202}Hg, and ^{204}Hg are shown in figure 5.2.2b. For the odd-numbered isotopes, with nuclear spins of 1/2, 3/2, and 5/2, each of the hyperfine components is split in a weak magnetic field to yield a multiplicity of lines (figure 5.2.3). However, in a strong field the coupling between the atomic angular momentum and the nuclear spin breaks down and the Zeeman pattern tends to revert to the simple triplet form. In the case of even-numbered isotopes, all three components are seen when viewed transversely to the magnetic field. The central π component is polarised parallel to the magnetic field and the displaced components, σ^+ and σ^-, are polarised perpendicular to the field (figure 5.2.2b).

Differential measurements of absorption during the periods of 'field' and 'no field' yield a direct value of the mercury vapour concentration. The method is sensitive, with $\epsilon = 4.2 \times 10^{-6}$ (Uthe et al, 1970). An increase in sensitivity is obtained when stannous chloride is added to the solution (Poluektov & Vitkun, 1963; Hingle et al, 1967). Combustion in a stream of air or oxygen will dissociate mercury compounds, and the mercury vapour thereby produced can be swept, by the gas, into the sample chamber of an atomic absorption spectrometer. Methods have been described for minerals (Kusnetsov et al, 1964), soil and stream sediment samples (Mitra & Webb, 1967; Mitra, 1970), air (Barringer, 1966), organic materials (Woodriff & Schrader, 1971), and other materials (Razumov & Utkina, 1969). The insertion of gold or platinum, coated on asbestos or in foil form, into the gas stream traps mercury. This mercury can subsequently be vaporized by heating, and measured without interference. Methods have been given for soils and rocks (James & Webb, 1963; Vaughn & McCarthy, 1964; Friedrich & Julms, 1969; Ward et al, 1969; Jonasson et al, 1973), and for air, blood, urine, fish, etc (Williston, 1968; Thilliez, 1968; Kalb, 1970).

Wet ashing methods are commonly used for generating mercury vapour by reduction. Most procedures use $SnCl_2$ and NH_2OH at high acidity, as described by Hatch and Ott (1968). Ascorbic acid is also effective (Shimomura et al, 1970). A low-temperature digestion apparatus (e.g. for fish samples) and method have been described by Uthe et al (1970) and Thorpe (1971). Krause et al (1971) give methods for urine, blood, water, and air analysis using H_2SO_4/HNO_3 digestion. The Dow Chemical Company has published details of methods for water, brines, caustics, fish, sludges, mud, hydrogen, and air (Anon, 1970).

A more advanced technique makes use of the different planes of polarisation of the centre and split-emission components. Using an electro-optical modulator and analyser, the two components can be sequentially routed through the spectrometer. The principle and the instrumentation of the supersensitive Zeeman spectrometer for trace mercury detection has been discussed in detail by Hadeishi and McLaughlin (1966). In this system, the Zeeman effect and a source containing ^{196}Hg are used to give two standard beams. One measures the total absorption and the other measures only the non-mercury absorption. Scintrex Ltd of Ontario have made a sensitive Zeeman magnetic system with electro-optical modulation for mercury detection during geological and pollution field studies.

Flame methods are specific for mercury, but some interference from cobalt, having a line at 253.649nm, has been reported by Manning and Fernandez (1968). Cold vapour methods, although free from metallic interference, may give errors if oxides of nitrogen, incompletely oxidised organic vapours, or smoke, having absorption lines at 253.7nm, are present (Mitra, 1973). Barringer (1966) has designed an instrument in which the absorption at the central part of the 253.7nm line is corrected, for the absorption which is not due to mercury, by subtracting absorption from the wings of the pressure-broadened source. A somewhat similar system has been described by Ling (1968).

Magos (1971) has described a method for the determination of inorganic mercury and methylmercury in biological samples. The method is based upon the knowledge that the rate of reduction of methylmercury or other organomercurials by tin^{II} chloride can be made identical to that of inorganic mercury if the amount of tin^{II} chloride is above a certain level and a cadmium salt is added to the reaction mixture. Thus, either inorganic mercury plus methylmercury, or inorganic mercury alone, can be released from the sample and determined using atomic absorption. Furthermore, it is possible to release inorganic mercury first, and then methylmercury, from the same sample. The details of his procedure are given below in order to illustrate the steps involved.

The apparatus used by Magos was a mercury vapour concentration meter manufactured by the Hendrey Relays Division of Columbia Industrial Development Ltd (Slough, UK) in which a water-pump was substituted for the original fan:

> The inlet of the gas cell was connected in succession, using plastic tubing, with two midget impingers of 30ml volume and a 200ml quickfit test tube fitted with a Dreschel bottle head, the inlet of which was converted into a thistle funnel. The first midget impinger was left empty and served as a liquid trap. The second impinger contained 10ml of distilled water and was immersed in melting ice so as to act as a water vapour absorber. The quickfit test tube was the reaction vessel.

Preparation of samples:

Homogenates were prepared in 1% solution.

Reagents:

Tin (II) chloride in 100mg portions.

Tin (II) chloride - cadmium chloride reagent: 25g of the tin chloride and 5g of the cadmium chloride were mixed with distilled water and heated to the boiling point. The volume was made up to 50ml with distilled water after cooling.

Cysteine hydrochloride - 1%w/v solution.

Method:

A small amount of homogenate was pipetted into the reaction vessel and 1ml of cysteine solution was added. The volume was then made up to 21.23ml with 1% saline solution, and 10ml of 16N H_2SO_4 was added. Next, either 1ml of tin (II) chloride - cadmium chloride reagent (method 1) or 100mg of tin (II) chloride (method 2) were added to the sample. The Dreschel head was connected to the reaction vessel and 20ml of 45% sodium hydroxide solution were added through the thistle funnel. The peak deflection was read and the concentration was calculated by the use of a standard.

Results:

Method 1 released all of the mercury from the sample, irrespective of the presence of a carbon-mercury bond. Method 2 released inorganic mercury at a rapid rate, and mercury from the methyl bond at a very slow rate. Using methods 1 and 2, determination of the total mercury content or the inorganic mercury content is possible. The difference between the values obtained using method 2 and method 1 gives the methylmercury content of the sample. The combination of methods 1 and 2 also makes possible the separate determination of inorganic mercury and organomercury in the same sample.

The method developed by Krause et al (1971) has the advantage that it requires very small amounts of sample (1ml or less). The procedure accurately determines the concentration of mercury in urine and blood specimens as well as in effluent water and working atmospheres where mercury is used. The method involves digestion of the specimens in order to release mercury vapour which is then reduced by stannous chloride and, after the removal of moisture, is monitored using an optical-cell mercury meter.

Apparatus:

This (figure 5.3) consists of an all-glass 100ml reaction vessel with a receiving funnel insert, connected to a water scrubber by means of a three-way stop-cock. From the scrubber, glass fittings connect to the optical cell. A disc of No.1 Whatman filter paper is inserted and sealed using an O-ring joint. This removes any droplets of moisture. Aspiration is made through coarse frits positioned near to the bottom of the flasks, as shown. A flow meter, calibrated for flow rates of between 1 and 4 l/min, is inserted downstream of the optical cell and upstream of the vacuum pump.

The Beckman model K-23 double-scale mercury vapour test meter used in these experiments was modified by removing the grille and mounting and positioning an all-glass optical cell. This was 22.5cm in length by 4cm in diameter, fitted with inlet and outlet parts, and with quartz glass windows fused into the ends.

Procedure:

Urine or water samples: To 1ml of urine in a beaker was added 5ml of concentrated HNO_3. This was allowed to stand for 2min before adding antifoaming reagent. The reactant was transferred quantitatively to the reaction vessel and 1ml of stannous chloride solution was added. The three-way stopcock was turned from the by-pass position to the aspiration position. The peak response was read on the meter and compared with a standard curve.

Blood samples:

To 1ml of heparinised or citrated blood in a beaker were added

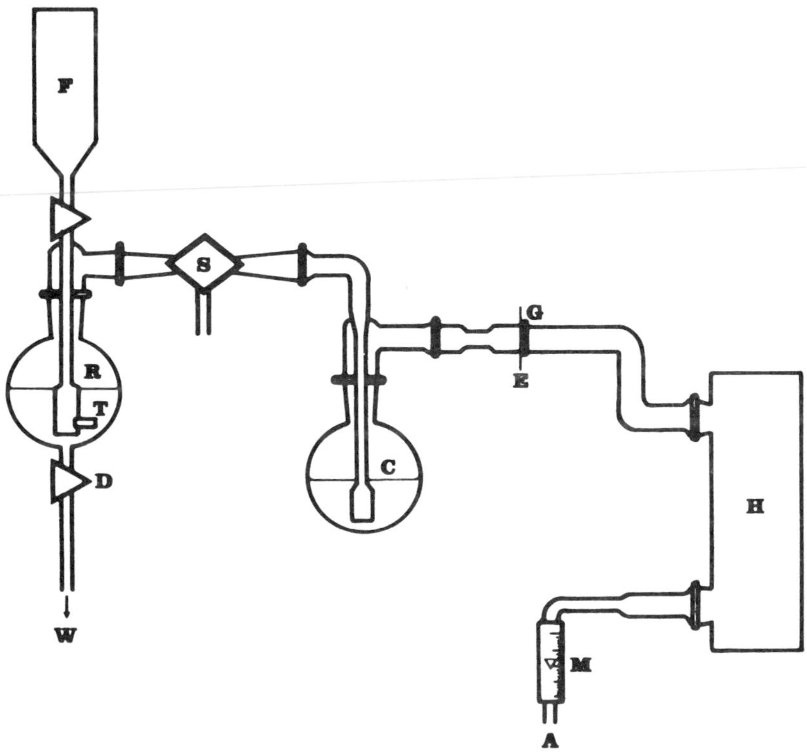

Figure 5.3 Urinary Mercury Analyser Connected to Optical Cell. (A: aspiration, C: water scrubber, D: drainage stopcock, E: Whatman filter paper, F: reaction funnel, G: O-ring, H: optical cell, M: rotometer, R: reaction vessel, S: three-way stopcock, T: frit, W: waste)

5ml of concentrated HNO_3 and 5ml of concentrated H_2SO_4. This was allowed to digest for 20min over a low flame or until an amber colour was fully developed. From this mixture, a 1ml sample was taken, mixed with 5ml of HNO_3, antifoaming reagent, and 1ml of stannous chloride solution. This was then transferred to the reaction vessel and processed as in the case of urine samples.

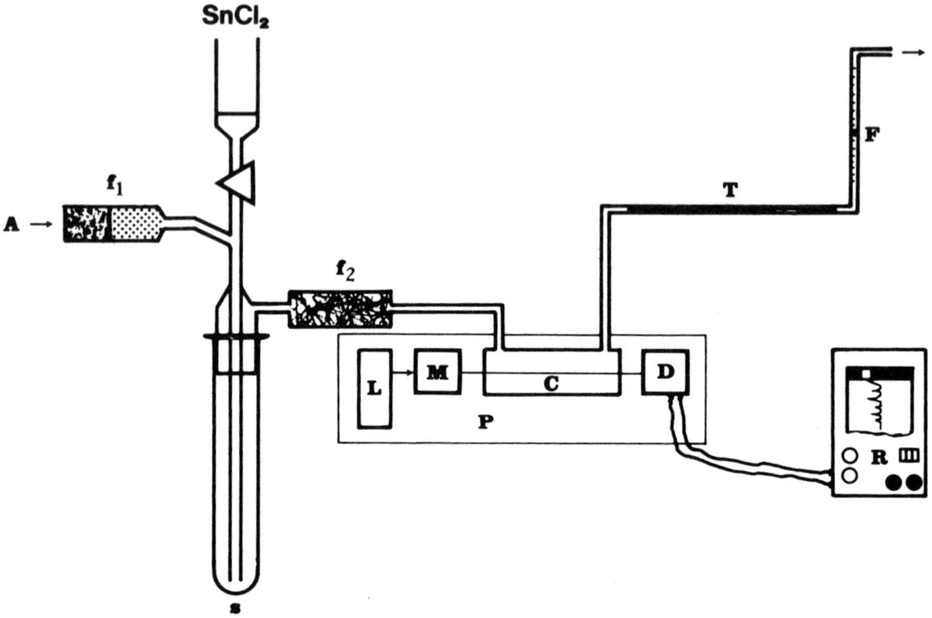

Figure 5.4 Apparatus for Mercury Determination (Lindstedt, 1970; p265). (A: air inlet, C: 10cm gas cell with quartz windows, D: detector, F: flow-meter, f_1: activated carbon and gold filter, f_2: cotton-wool filter, L: mercury lamp, M: monochromator, P: spectrophotometer section, R: recorder, S: suction, s: sample, T: capillary)

A method using direct injection into a flame was described by Lindstrom (1959) and by Monkman et al (1956) for urine, etc. A concentrate obtained by APDC/MIBK extraction was used by Willis (1962) for urine. Messman and Smith (1970) decomposed an APDC/MIBK extract from urine by using the boat technique.

Another atomic absorption method has been suggested by Robinson et al (1972) who directly determined mercury in air, using the 184.9nm resonance line:

> The air sample is passed through a column of carbon pieces at about 1350°C. As a result, O is converted into CO, and metallic compounds in the air are reduced to the atomic state. The

products are passed into a silica tube heated to about 900°C, and radiation from a suitable source is passed through the tube into a monochromator (in which the mercury is replaced by Ar), and then to a measuring system.

Flameless atomic absorption methods have been widely used, with modifications in sample treatment, for the determination of trace amounts of mercury in geological samples (Friberg & Vostal, 1972; Flanagan et al, 1982), vegetation (McHugh & Turner, 1975), natural waters and soils.
The procedure for cold vapour AAS analyses of liquid samples (urine, etc), used by Lindstedt (1970), is given below as a guide:

Pipette 2ml of solution (urine or mercury standard), having a mercury content in the range: 0 - 0.2 μg/ml, into a series of tubes with ground glass joints. Into each, mix 0.4ml of concentrated H_2SO_4 and cool in tap water. Add and mix, 3ml of $KMnO_4$ solution (60g/l) and leave the tubes loosely stoppered overnight at room temperature. After at least 12 hours, add 0.6ml of hydroxylammonium chloride (20g/100ml) in order to reduce the excess $KMnO_4$.
Add 5ml of stannous chloride solution (20g $SnCl_2.2H_2O$ in 100ml of 6M HCl) to the first tube and immediately fit the tube to the apparatus as shown in figure 5.4, with the air flow vented to the atmosphere. Turn the stopcock so as to pass air through the sample and into the absorption cell at a rate of 3 litres per minute. Record the mercury absorption peak at 253.6nm. Replace the tube containing the sample with one containing 20ml of distilled water and flush air through the system until the baseline absorbance returns to its initial value. Continue with the remaining samples and standards.

E. NAA

This is a highly specific and sensitive method for the determination of mercury. The method is also suitable for the analysis of a large number of samples. In order to use it, access to both an atomic reactor with a thermal neutron flux of between 10^{11} and 10^{13}/cm^{2}s, and to a counting device with a pulse height analyser, is required. When mercury with the natural isotopic composition is irradiated with thermal neutrons, five nucleides are produced, but only ^{197}Hg, with a half-life of 65h and ^{203}Hg with a half-life of 47 days, are normally convenient for analytical use. Usually, the 77keV gamma-rays and 68keV X-rays from ^{197}Hg and the 279keV gamma-rays from ^{203}Hg are used for counting.

Non-destructive methods for the neutron activation analysis of biological and environmental samples, and for rocks and minerals, have been described by Filby et al (1970) and by Laul et al (1970), respectively.

Digestion methods have been used for biological materials (Pillay et al, 1971), urine (Ishida et al, 1970), human hair (Krause et al, 1971), plant tissue (Haller et al, 1968; Silva & Lima, 1970), and rocks (Alian & Shabana, 1967; Kiesl, 1969). The use of ion exchange resins for the separation of mercury in this analysis is described by Becknell et al (1971), Samsahl (1967), and Jones et al (1971). Kosta and Byrne (1969) ignited irradiated material in oxygen and trapped mercury in Se-impregnated paper.

In neutron activation analysis using reactor neutrons, it is highly desirable to have the samples in a compacted form as well as being free of excessive amounts of moisture. Processes such as oven drying, freeze drying, ashing, etc, with precautions being taken against any loss of mercury, are often used to prepare samples. Mercury in biological tissues can be stabilised against loss during freeze drying by exposure to high doses of high-energy nuclear radiation.

The method adopted by Pillay et al (1971) is given in detail below, as an illustration of the procedure adopted for environmental sample analysis.

The sample, encapsulated in a heavy-duty polyethylene container together with a mercury standard (contained in a quartz vial), was irradiated using a thermal neutron flux of about 5×10^{12}/cm^{2}s for 2h. Samples were allowed to decay for at least 1h before processing in order to permit the short-lived activity of the matrix to decay.

Wet ashing and/or extractive digestion:

The irradiated sample contained in the polyethylene bag was removed and inserted into a 200ml distillation flask containing an accurately known amount of mercury carrier (50mg Hg per ml as Hg^{2+}) and 10ml of concentrated HNO_3. The condensing apparatus was fixed to the distillation flask and 5ml of concentrated H_2SO_4 and 5ml of 70% perchloric acid were added through the top of the condenser. The contents of the flask were first simmered and then slowly heated to 150°C before cooling. If traces of undissolved or charred tissue and oily suspensions remained, the refluxing was repeated.

Separation of mercury:

The contents of the flask were transferred to a beaker, and excess acid was neutralised using NH_4OH. The solution was then acidified (to about 2M) with HCl. The solution was warmed and treated with H_2S gas until precipitation was complete. The sulphide precipitates were first washed with warm 2M HCl in order to remove contaminants. After washing with warm water, the precipitates were mixed with 5ml of ammonium polysulphide and 1ml of 2M NaOH and kept in a water bath in order to permit the dissolution of antimony, arsenic, and tin sulphides. The precipitates were then washed with warm water, followed by 3M HNO_3, in order to remove certain trace element contaminants. The mercuric sulphide precipitate was then mixed with 3 to 10 drops of freshly prepared aqua regia and kept warm on a water bath for 10 to 15min. The mercury salts were then diluted with about 30ml of water and warmed in order to coagulate the sulphur. The solution was filtered into a beaker through a coarse filter paper.

Electrolysis:

The deposition of mercury was carried out using platinum as the anode and gold foil (2 x 2 x 0.02cm) as the cathode. The gold foils were properly marked and weighed before electrolysis. Duplicate samples of the standard were also electrolysed together with the samples. The electrolytic cells were connected in parallel and a DC potential of about 4.5V was applied. The total current flowing through 16 cells was less than 0.8A.

Electrolysis under these conditions permitted extremely uniform deposition.

After electrolysis, the cathodes were identified and rinsed with deionised water and ethyl alcohol. The excess alcohol was wiped off using absorbent tissue paper. The foils were then either air-dried or oven-dried at 60°C.

Counting:

The gamma-ray and X-ray emissions from ^{197}Hg and ^{197m}Hg were counted using a thin (0.6 x 5cm) sodium iodide detector with a beryllium window and a 400-channel pulse height analyser.

This method has been successfully used for the determination of Hg in coal samples, air particulates, and a variety of foodstuffs.

An outline for the determination of mercury, together with cadmium, tellurium and bismuth, in some major rock-forming minerals has been discussed by Marowsky and Wedepohl (1971). The sample preparation, irradiation, radiochemical separation, and radiometric procedures used for silicate analysis have been discussed. A batch of five powdered rock samples, (1 - 2g each), together with a flux monitor consisting of 1 μg Cd, Hg, Ti, and Bi were irradiated for 3 days using a thermal neutron flux of about $6 \times 10^{13}/cm^2s$ in the FRJ-1 reactor of the Kernforschungsanlage Jülich (Germany). After dissolution of the samples in the presence of inactive carriers of each of the four elements, and suitable hold-back carriers for some interfering nucleides, the bulk activity of the matrix was removed using an anion exchange step. The isotopes used were:

Element	Isotope	Half-life	Energy (gamma-ray)
Cd	^{115}Cd	53.5h	0.0530MeV
Hg	^{203}Hg	47.0d	0.279Mev

Radiochemical procedures were selected so as to increase the chemical yield, which varied from 40 to 80%. The Cd and Hg were purified for gamma-spectrometry using a

$28cm^3$ Li-drifted Ge detector and a 4000 channel analyser.
The Hg and Te could be detected at the 0.1ppb level. The limits of detection of the above-mentioned elements have been discussed elsewhere (Marowsky, 1971). The relative standard deviations given by the counting statistics were always less then 3% at higher concentrations*.
Neutron activation analysis has been successfully used for the determination of mercury in foodstuffs (Harry et al, 1972; Tejam & Halder, 1975), biological tissues (Heitzmann & Simpson, 1972), rocks (Dissanayake & Vincent, 1975).
The reliability of mercury analysis in environmental samples has been discussed by Heinonen et al (1972), and later by the US Food and Drugs Administration (Schafer et al, 1975). In deciding real levels, the importance of statistical evaluation of the results has been emphasized by many workers (Abbey, 1981). Inter-laboratory investigations of the determination of mercury in dried milk powder revealed a very wide range of values, giving a mean value of 35.5ng/g, with a standard deviation of 124%, while the real level was later concluded to be about 0.5ng/g (Tölg, 1977).
In 1966, the International Atomic Energy Agency organised a series of intercomparisons for the determination of mercury in biological materials by neutron activation and other techniques (Tugsavul et al, 1970). The IAEA, under its Analytical Quality Control scheme, provides materials for intercomparison, as well as standard reference materials.

* Using this procedure, Marowsky and Wedepohl (1971) analysed common magmatic rocks, including the international standards and some sedimentary and igneous rocks. In their work, they also compared the distribution of Cd, Hg, Ti, and Bi in metamorphic rocks of progressive metamorphism, and the distribution between sedimentary and igneous rocks.

REFERENCES

- Abbey, S. (1981) - Analytical Chemistry, 54(4), 5294
- Ahmed, R., Valenta, P., and Nurnberg, H.W. (1981) - Mikrochimica Acta, 171-84
- Akagi, H., Mortimer, D.C., and Miller, D.R. (1979) - Bulletin of Environmental Contamination and Toxicology, 23, 372-6
- Alian, A. and Shabana, R. (1967) - Microchemical Journal, 12, 427
- Andrews, R.W., Larochelle, J.H., and Johnson, D.C. (1976) - Analytical Chemistry, 48, 212-4
- Anon (1970) - Determination of Mercury by the Atomic Absorption Spectrophotometric Method - Applicable to Water, Brines, Caustics, Fish, Sludges, Mud, Hydrogen, and Air, Dow Chemical Company, Midland, Michigan
- Barringer, A.R. (1966) - Transactions of the Institute of Mining and Metallurgy, B75, 120
- Becknell, D.E., March, R.H., and Allie, W. (1971) - Analytical Chemistry, 43, 1230
- Ben-Bassat, D. and Mayer, A.M. (1977) - Physiologia, Pl., 40, 157-62
- Cholak, J. and Hubbard, D.M. (1946) - Industrial and Engineering Chemistry (Anal. Ed.), 1, 149
- Clarkson, T.W. and Greenwood, M.R. (1970) - Analytical Biochemistry, 37, 237-43
- De Filippis, L.F. (1978) - Zeitschrift für Pflanzenphysiologie, 86, 339-52
- Dissanayake, C.B. and Vincent, E.A. (1975) - Mineralogical Magazine, 40, 33
- Dolar, S.G., Kenny, D.R., and Chesters, G. (1971) - Environmental Letters, 1, 191-8
- Ehmann, W.D. and Lovering, J.F. (1967) - Geochimica et Cosmochimica Acta, 31, 357
- Fang, S.C. (1973) - Archives of Environmental Contamination and Toxicology, 1, 18-26
- Filby, R.H., Davis, A.I., Shah, K.R., and Haller, W.A. (1970) - Mikrochimica Acta, 1130

- Flanagan, F.J., Moore, R., and Aruscavage, P.J. (1982) - Geostandards Newsletter, 6(1), 25-46

- Friberg, L. and Vostal, J. (1972) - in Mercury in the Environment (Edited by L.Friberg and J.Vostal), CRC Press, Cleveland, 215

- Friedrich, G. and Julms, M. (1969) - Erzmetall, 22, 74

- Gage, J.C. (1961) - Analyst, 86, 457

- Gay, D.D. (1976) - Methylmercury: Formation in Plant Tissues, US Environmental Protection Agency, Las Vegas, Nevada 89114, PB - 255096

- Gustavsson, I., Golimowski, J. (1981) - Science of the Total Environment, 22, 85-7

- Hadeishi, T. and McLaughlan, R.D. (1966) - Science, 174, 404-7

- Haller, W.A., Ranatcelli, L.A., and Cooper, J.A. (1968) - Journal of Agricultural and Food Chemistry, 16, 1036

- Harry, R.L., Gills, T.E., and Lallur, P.D. (1972) - Analytical Chemistry, 44[7], 114

- Hatch, W.R. and Ott, W.L. (1968) - Analytical Chemistry, 40, 3085

- Heinonen, J., Marten, D., and Suscheey, O. (1972) - Mercury Contamination in Man and his Environment, IAEA., Technical Report Series No. 137, 137-41

- Heitzmann, M.W. and Simpson, R.E. (1972) - Journal of the Association of Official Analytical Chemists, 55, 960

- Hingle, D.N., Kinkbright, G.F., and West, T.S. (1967) - Analyst, 92, 759

- Horwitz, W. (Editor) (1965) - Methods of Analysis of the Association of Official Agricultural Chemists, AOAC (USA), 375

- Iarayan, V.M., Ovespyan, E.N., and Karimyan, N.S. (1973) - Izvestiya Vuisshikh Uchebnuikh Zavedeny Khimiya Tekhnologiya, 16, 358

- Ishida, K., Kawamura, S., and Izawa, M. (1970) - Analytica Chimica Acta, 50, 351

- James, C.H. and Webb, J.S. (1963) - Transactions of the Institute of Mining and Metallurgy, 73, 633

- Jensen, K.D. and Carlsen, V. (1978) - Journal of Radioanalytical Chemistry, 47, 121-34

- Jonasson, I.R., Lynch, J.J., and Trip, L.J. (1973) - Geological Survey of Canada Paper, 73-21, 22

- Jones, J.D., Rottschafer, J.M., Mark, H.B., Paulsen, K.E., and Patriarche, G.J. (1971) - Mikrochimica Acta, 399

- Jones, P. and Nickless, G. (1974) - Journal of Chromatography, 89, 201-8
- Kalb, G.W. (1970) - Atomic Abstract News Letters, 9, 84
- Kamath et al (1975) - Personal communication
- Kasper, J. and Kral, R. (1958) - Sbornik Vysokeskoly Khimicheskikh Tekhnologii V. Praze, Otdelnuy Fakultat Anorganicheskikh i Organicheskikh Tekhnologii, 281
- Kiesl, W. (1969) - National Bureau of Standards (US) Special Publication 302
- Kosta, L. and Byrne, A.R. (1969) - Talanta, 16, 1297
- Kothny, E.L. (1970) - American Industrial Hygiene Association Journal, 31, 466
- Krause, L.A., Henderson, R., Shotwell, H.P., and Culp, D.A. (1971) - American Industrial Hygiene Association Journal, 32, 331
- Kudo, A., Akagi, H., Mortimer, D.C., and Miller, D.R. (1977a) - Environmental Science and Technology, 11, 907
- Kudo, A., Akagi, H., Mortimer, D.C., and Miller, D.R. (1977b) - Nature, 270, 419-20
- Kuznetsov, N. and Chabovsku, L.P. (1964) - Uchebnik Zapadnogo Tsentra Nauchno-Issledovatelski Institut Olovyan Promuishlennost, 75
- Laul, J.C., Case, D.R., Wechter, M., Schmidt-Bleck, F., and Lipschutz, M.S. (1970) - Journal of Radioanalytical Chemistry, 4, 241
- Lindstedt, G. (1970) - Analyst, 95, 264-71
- Lindstrom, O. (1959) - Analytical Chemistry, 31, 461
- Ling, C. (1968) - Analytical Chemistry, 40, 1876
- Litman, R., Williams, E.T., and Finston, H.L. (1977) - Analytical Chemistry, 49, 983-7
- Magos, L. (1971) - Analyst, 96, 847
- Manning, D.C. and Fernandez, F. (1968) - Atomic Abstract News Letters, 7, 24
- Marowsky, G. (1971) - Zeitschrift für Analytische Chemie, 253, 267-71
- Marowsky, G. and Wedepohl, K.H. (1971) - Geochimica et Cosmochimica Acta, 35, 1255-67
- Mazumdar, M. and Shome, S.C. (1971) - Analytica Chimica Acta, 56, 149
- McHugh, J.B. and Turner, R.L. (1975) - US Geological Survey Bulletin, 1408, 21

- McNerney, J.T. and Buseck, P.R. (1973) - Economic Geology, 68[8], 1313-20
- Messman, B.B. and Smith, B.S. (1970) - Atomic Abstract News Letters, 9, 81
- Mitra, S. (1970) - Proceedings of the Indian National Science Academy, 36A, 4, 191-210
- Mitra, S. (1973) - Mercurometric Survey for Mineral Exploration with a Case Study (abstract) Indo-Soviet Symposium on Recent Trends in the Exploration of Minerals, Oil and Groundwater, New Delhi, October, 1973
- Mitra, S. and Ghosh Dastidar, P. (1976) - Science and Culture, 42, 9, 443-51
- Mitra, S. and Webb, J.S. (1967) - Transactions of the Institute of Mining and Metallurgy, 76B, 180
- Monkman, J.L., Maffeett. P.A., and Doherty, T.F. (1956) - American Industrial Hygiene Association Journal, 17, 418
- Olafsson, J. (1974) - Analytica Chimica Acta, 68, 207-11
- Paul, M., Tibbetts, F.E., Zarnegar, P., and Fisher, G.B. (1973) - Journal of Chromatography, 87, 215
- Pawlaczyk, J. and Sierzant, M. (1974) - Farmacja Polska, 30, 37
- Pillay, K.K.S., Thomas, C.C., Sondel, J.A., and Hyche, C.M. (1971) - Analytical Chemistry, 43, 1419-25
- Poluektov, N.S. and Vitkun, R.A. (1963) - Zhurnal Analytischkoi Khimie, 18, 37
- Popea, F. and Jemaneanu, M. (1960) - Academia Republicii Socialiste Romania Studisi Cercetari de Chimica, 8, 607
- Razumov, V.A. and Utkina, T.P. (1969) - Spektroskopiya Trudy Sibirskogo Soveshchaniya, 1065, 291
- Robinson, J.W., Slevin, P.J., Hindman, G.D., and Wolcott, D.K. (1972) - Analytica Chimica Acta, 61(3), 431
- Rodriguez-Vazquez, J.A. (1978) - Talanta, 25, 299
- Samsahl, K. (1967) - Analytical Chemistry, 39, 1380
- Sawaya, T., Ishi, H., and Odashima, T. (1973) - Japan Analyst, 22, 318
- Schafer, M.L., Rhea, U., Peeler, J.T., Hamilton, C.H., and Campbell, J.E. (1975) - Journal of Agriculture and Food Chemistry, 23, 1079
- Shimomura, S., Fukamoto, Y., Hashimoto, M., and Tanase, Y. (1970) - Bunseki Kagaku, 19, 1296

- Silva, C.M. and Lima, F.W. (1970) - Revista Brasileira de Technologia, 67 in Analytical Abstracts, 21, 2971 (1971)

- Sipos, L., Golimowski, J., Valenta, P., and Nürnberg, H.W. (1979) - Fresenius Zeitschrift für Analytische Chemie, 298, 1-5

- Stary, J. and Kratzer, K. (1980) - Radiochemistry and Radioanalytical Letters, 44, 37-40

- Stary, J., Kratzer, K., Havlik, B., Prasilova, J., and Hanusova, J. (1980) - Int. J. Environ. Anal. Chem., 8, 189-95

- Stoeppler, M., Valenta, P., and Nürnberg, H.W. (1979) - Fresenius Zeitschrift für Analytische Chemie, 297, 22-34

- Supronovich, I.B. (1936) - Journal of General Chemistry of the USSR, 8, 839

- Tejam, B.M. and Halder, B.C. (1975) - Indian Journal of Environmental Health, 17[1], 9

- Thabet, S. and Tabibian, O. (1966) - Analytica Chimica Acta, 34, 241

- Thilliez, G. (1968) - Chimie Analytique, 50, 226

- Thorpe, V.A. (1971) - Journal of the Association of Official Agricultural Chemists, 54, 206

- Tölg, G. (1977) - Zeitschrift für Analytische Chemie, 283, 257

- Tungsavul, A., Merren, D., and Suschney, O. (1970) - The Reliability of Low-Level Radiochemical Analysis: Results of Intercomparisons Organised by the IAEA, 1966-69, IAEA Report, 30th March, 1970

- Ure, A.M. and Shand, C.A. (1974) - Analytica Chimica Acta, 72, 63-77

- Uthe, J.F., Armstrong, F.A.J., and Stainton, M.P. (1970) - Journal of the Fisheries Research Board of Canada, 27, 805

- Vaughn. W.W. and McCarthy, J.H. (1964) - US Geological Survey Professional Paper No. 501-D, 123

- Ward, F.N. (1970) - in Mercury in the Environment, USGS Professional Paper 713, 46-9

- Ward, F.N. (1975) - USGS Bulletin, 1408

- Ward, F.N., Nakagawa, H.M., Harms, T.S., et al (1969) - US Geological Survey Bulletin, 1152, 100

- Westöö, G. (1966) - Acta Chimica Scandinavica, 20, 2131

- Westöö, G. (1967) - Acta Chimica Scandinavica, 21, 1790
- Willis, J.B. (1962) - Analytical Chemistry, 34, 615
- Williston, S.H. (1968) - Journal of Geophysical Research, 73, 7051
- Windom, H., Gardner, W., Stephens, J., and Taylor, F. (1976) - Estuarine and Coastal Marine Science, 4, 579-83
- Woodriff, R. and Schrader, D. (1971) - Analytical Chemistry, 43, 1918
- Zarnegar, P. and Mushak, P. (1974) - Analytica Chimica Acta, 69, 389
- Zmijewska, W. (1977) - Journal of Radioanalytical Chemistry, 35, 389-418

CHAPTER VI
MERCURY IN MAMMALS

INTRODUCTION

Mercury in any form is toxic to man and other living things. In non-occupational exposure to mercury, the routes of entry into the human system are quite different for the various forms of mercury. The organic forms are more likely to be taken in from food and water, whereas elemental mercury is mainly absorbed by inhalation of the vapour. Mercury levels are very high in the members of those species which come into contact with mercury fungicides, particularly seed-eating birds and mammals, predators which prey on these animals, and man (figure 6.1).

Methylmercury is deposited in (and irreversibly destroys) brain and nerve cells (Mottett, 1973; WHO, 1976). The inorganic salts of mercury may be inhaled or ingested via environmental exposure. Clinical and sub-clinical effects of mercury absorption have been found in dentists (Joselow et al, 1968), laboratory workers (Goldwater et al, 1956), industrial workers (Battigelli, 1960), and miners (Rossi et al, 1976).

The lungs of slaughtered goats and lambs in the mining districts around Asansol-Dhanbad, and in the industrialised zones around Calcutta, contain 1401 to 9099ppm of mercury (ITRC, 1981). In areas of atmospheric mercury pollution, sub-clinical effects such as micromercurialism may occur with concentration in the tissues (Skerfving & Vostal, 1972; Friberg & Nordberg, 1972). Organic mercury in the human system comes mainly from fish, while inorganic mercury comes from the atmosphere, drinking water, the breakdown of phenylmercury pesticides and fungicides in agriculture, dental amalgams, and also from the breakdown of methylmercury. The latter occurs at the rate of 1% per day in the body (Clarkson, 1977). The body slowly discharges its mercury burden via the urine, faeces, hair, sweat, and breath. Mercury is excreted from the blood, into hair, in both organic and inorganic forms (Airey, 1983). Human blood and hair have been used to monitor mercury intake in toxicological studies (Friberg & Vostal, 1972).

In this chapter, the routes of entry and the content of mercury in the human and animal system are considered.

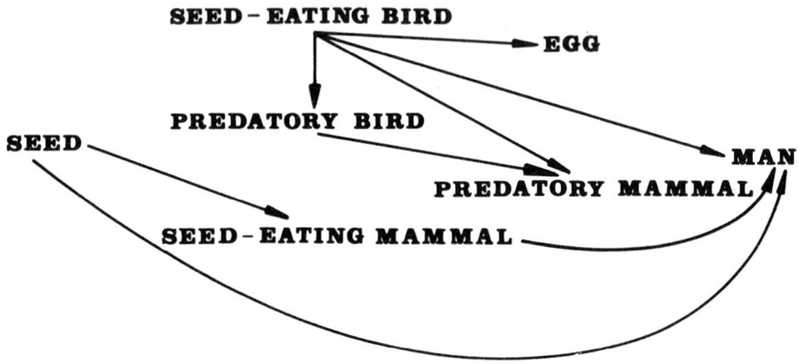

Figure 6.1 Mercury in Food Chains

Metallic mercury and other mercury compounds enter the body via the gastro-intestinal tract, respiratory system, mucous membranes, and skin. The rate of absorption depends upon the nature of the compound and upon other factors. The accumulation of mercury is markedly dependent upon the continuity of exposure. The clinical manifestations and the treatment of poisoning have been discussed elsewhere (Mitra & Ghosh Dastidar, 1976).

FUNGICIDE SOURCE

The use of mercury-based pesticides can contribute significantly to environmental pollution. Seed dressing can introduce as much as 10g of mercury per hectare of soil. Grain treated with mercury may be consumed by wild animals, which can accumulate significant amounts of mercury in their bodies. Mercury fungicides were banned in Sweden in 1966, when their role in the ecological cycle of seed-eating birds and their predators was proved.

Most mammals ingest mercury fungicides directly by eating spray-treated vegetables. Man can also absorb mercury fungicides via the lungs, skin, or stomach (Friberg &

Vostal, 1972) and these are equilibrated with selective proteins in the body. Factory workers and farmers inhale mercury from spray dusts or while handling treated seeds. The accumulation of mercury in man is strongly dependent upon the continuity of exposure. However, the accumulation of mercury in the edible portion of corn is very limited and mercury fungicides in agriculture do not constitute any serious health hazard for man. On the other hand, farmers may occasionally use leftover treated seed as animal feed and thereby lead to increased mercury levels in the flesh of animals, and also in the eggs of hens. In New Mexico, four members of a family were seriously poisoned due to eating the flesh of a hog which had been fed with alkylmercury-treated seed (Curley et al, 1971).

However, deaths caused by the direct human ingestion of treated cereal seeds have been reported from various places. Engleson and Herner (1952) reported the death of three persons who were poisoned by eating porridge made from cereal seed. In northern Iraq, many hundreds of persons succumbed in 1956 due to eating home-baked bread made using ethylmercury-treated cereal seed. In Pakistan, more than one hundred persons were poisoned by the ingestion of bread made from wheat seed treated with ethylmercury and methylmercury compounds (Haq, 1963).

Small rodents in Sweden contain elevated mercury levels due to having eaten fungicide-treated seed (Lihnell & Stenmark, 1967). The death of domestic animals due to the consumption of treated seed (Curley et al, 1971), and mercurialism in some predators (Berg et al, 1966) are well-known.

INDUSTRIAL EXPOSURE

A large proportion of the mercury found in the environment is derived from industrially produced mercury, and amounts to approximately 10000 tons per year. In the northern hemisphere, higher concentrations of mercury have been observed in air, sea-water, and in the hair of the inhabitants. The industries which are mainly responsible for the wide dispersion of mercury are those which involve the burning of fossil fuels, compost incineration, the mining and extraction of mercury from cinnabar, chlor-alkali manufacture, paper pulp manufacture, paint manufacture, fungicide manufacture, electrical equipment manufacture, the repair of DC meters and pressure gauges, and amalgamation. Mercury droplets from broken thermometers can also be a source of exposure to mercury vapour in the homes or in hospitals.

Important sources of mercury are the burning of fossil fuels and ores. Although the concentration of mercury in such fuels is small, the latter are consumed at an enormous rate and therefore contribute a significant amount of mercury to the environment. The world-wide release of mercury from coal is of the order of 3000 tonnes per year.

The mercury content of ores (excluding mercury ores themselves) is probably higher than that of coal. However, the tonnage of ore mined annually is much lower than that of coal. The world-wide release of mercury from ore processing is more than 2000 tonnes per year.

If the average mercury content of crude oil is taken to be about 0.1ppm, then some 200 tons of mercury can be expected to be released annually into the atmosphere due to the combustion of petroleum alone.

Another common mercury hazard is its use in dentistry, where it is used to make mercury amalgams for fillings, as well as in drugs, disinfectants, and sterilising solutions.

The use of so-called 'grey powder' (chalk and mercury mixed in the ratio of 2 to 1) for latent fingerprint development has been held to be responsible for the occurrence of mercurialism in several detectives.

In earlier times, the mercury nitrate process for the felting of fur used in the manufacture of hats caused stomatitis (inflammation of the mouth), excessive salivation, tremors, and muscular atrophy.

Each year, about 20000 tons of mercury are precipitated at the earth's surface by rainfall, and some 70 million tons of mercury are present in the sea.

The extensive industrial and agricultural uses of mercury, and its subsequent environmental build-up, can cause dangerous levels to be reached in certain localised areas. In contaminated aquatic systems, mercury undergoes biotransformation into a very toxic monomethyl form which is effectively accumulated in fish and other aquatic organisms, or into dimethylmercury which, due to its volatility, is easily disseminated.

The tragic consequences of a combination of localised environmental pollution by mercury compounds and of a heavy consumption of sea-food by the people of the Minamata and Niigota regions of Japan are well known. However, an increasing world-wide shortage of protein food, especially in developing countries, will certainly increase fish consumption. Thus, this mercury hazard might become a more general environmental problem.

Table 6.1 Normal Mercury Contents of Various Organs of the Human Body

Organ	Content (mg/kg)	Reference
Hair	0.15 - 14.4	Kyle & Ghani, 1983
Hair	0.30 - 2.74	Flaschentrager, 1951
Hair	1.7 - 3.8	Airey, 1982
Skin	0.003 - 0.006	Forbes et al, 1954
Bone tissue	0.013	Stock, 1940
Bone marrow	0.019	Stock, 1940
Tooth enamel	0.1 - 2.15	Nixon et al, 1967
Muscle tissue	0.02 - 0.15	Forbes et al, 1954
Fatty tissue	0.03 - 0.08	Forbes et al, 1954
Nerve tissue	0.047	Forbes et al, 1954
Brain	$<$0.05 - 0.10	Joselow & Goldwater, 1967
Gastrointestinal tract	0.017	Forbes et al, 1954
Large intestine	0.05	Joselow & Goldwater, 1967
Small intestine	0.10	Joselow & Goldwater, 1967
Thymus	$<$0.5	Joselow & Goldwater, 1967
Pancreas	0.003 - 0.05	Stock, 1940
Lungs	0.017 - 0.10	Forbes et al, 1954
Heart	0.052	Forbes et al, 1954
Liver	0.01 - 0.12	Stock, 1940
Gallstones	0.011	Stock, 1940
Kidney	2.05	Griffith et al, 1954
Milk	0.12	Griffith at al, 1954
Milk	11.8 $\pm$ 8.3*	Kosta et al, 1983
	* μg/kg (dry weight)	

Because of the predominance of the neurological symptoms of Minamata disease, mercury (particularly in the form of alkylmercury) is considered to be a potent neurotoxin.

The daily absorption of mercury by human beings is of the order of 0.005 to 0.20mg. Small amounts of this element are normally present in all of the organs of the body. It appears that the kidneys contain the highest concentration of mercury. The typical mercury contents of various organs of the human body are given by table 6.1.

Table 6.2 Mercury Content* of the Kidneys, Liver, and Lungs of Swedish Workers (Brune et al, 1980)

Group	Parameter	Kidneys	No.	Liver	No.	Lungs	No.
Control	median	0.091	8	0.037	8	0.008	8
Control	range	0.054-0.027		0.019-0.14		0.001	
All	median	0.086	13	0.036	13	0.036	13
All	range	0.017-0.50		0.016-0.13		0.012-1.13	
I	median	0.01	9	0.034	9	0.031	9
I	range	0.017-0.50		0.016-0.10		0.018-1.13	
II	median	0.046	4	0.037	4	0.038	4
II	range	0.044-0.47		0.018-0.13		0.012-0.097	

* μg/g wet weight

SWEDISH EXAMPLE

The levels of mercury, and of 22 other toxic elements, in the kidneys, liver, and lungs of workers from a smelter and refinery (owned by the Boliden company of north Sweden), which processed non-ferrous concentrates of Cu, As, Au, Pb, Ni, Ag, and Zn, were determined by Brune et al (1980). The analyses (carried out using NAA and nucleides of ^{197}Hg and ^{203}Hg) indicated the levels of mercury in exposed workers as well as in members of a control group. The values were treated statistically using Wilcoxon's test. The control group consisted of 8 individuals with ages ranging from 43 to 72 years, who lived in the areas of Burtrask and Jorn (located about 50km from the factory region). The worker group consisted of 21 cases, and most had been exposed for between one and three decades. Their ages ranged from 55 to 76 years. The time between retirement and death ranged from 0 to 19 years. Multielemental analyses were performed on tissue from the kidneys, lungs, and liver. The workers were sub-divided into two groups on the basis of the time elapsed between retirement and death. Group I comprised the 0 to 1.5 year cases, and group II comprised the 2 to

19 year cases. The mercury contents are shown in table 6.2.

BRAIN AND OTHER TISSUES

Some workers have observed that mercury is localised in particular regions of the brains of milking animals. More mercury is accumulated in the grey matter of the brain as compared to the brain as a whole. Neutron activation analyses of human tissue revealed the presence of mercury in every part, but particularly in the cerebellum. The white matter contains the lowest percentage and this is probably because it lacks cell bodies for which mercury may have an affinity (Glomski et al, 1971).

Table 6.3 Mercury Concentration (ppb) in Human Tissues

Organ	Multiple Sclerosis	Mean	Standard Deviation
Frontal lobe	yes	3.3	1.8
Frontal lobe	no	2.9	1.7
Cerebellum	yes	3.5	0.8
Cerebellum	no	3.7	1.3
Spinal cord	yes	4.5	3.1
Spinal cord	no	4.5	2.4
Liver	yes	3.3	2.9
Liver	no	3.0	2.1
Muscle	yes	4.3	2.8
Muscle	no	2.3	0.7

Mercury determinations were carried out on samples obtained from 17 hospitals in British Columbia. The samples included post-mortem tissue from the frontal lobe, lateral lobe of the celebellum, spinal cord, liver, and muscles of patients with multiple sclerosis as well as from patients who were free of neurological disease (Warren et al, 1983). Some 5g of each frozen tissue sample were dissolved in 5ml of 20% tetramethyl ammonium hydroxide in alcohol. The results, obtained using the cold vapour absorption method, are shown in table 6.3. (This method has been described by S.J.Horsky in the Proceedings of the Spectroscopy Society - SSC - 10, Department of

Pathology, University of British Columbia, Vancouver BC).

DISTRIBUTION IN VARIOUS ORGANS

Methylmercury compounds are 100 times more soluble in lipids than in water and this fact accounts for the distribution of these compounds in the tissues.

Table 6.4 Concentration of Mercury in Blood (Goldwater et al, 1964)

Country	Samples	Approximate Level	Samples with less than 0.005ppm
Argentina	49	0 - 3.0	39
Chile	8	0.6 - 2.1	0
Columbia	5	0	5
Czechoslovakia	20	0 - 2.1	12
Egypt	14	0 - 0.9	13
England	30	0 - 2.7	29
Finland	46	0 - 7.5	32
Israel	38	0 - 3.0	31
Italy	27	0 - 3.0	22
Japan	40	0 - 2.7	32
Netherlands	43	0 - 2.1	39
Peru	32	0	7
Poland	85	0 - 31.5	59
Sweden	30	0 - 9.0	27
Yugoslavia	45	0 - 4.2	36
USA	98	0 - 7.0	71
Overall	609	0 - 31.5	454 (74.5%)

The distribution of mercury in individual organs and tissues has been studied by various workers. Ellis and Fang (1967) have found that, after the oral administration of mercury acetate or phenylmercury acetate to rats, the highest concentration is found in the kidneys, followed by the liver, lungs, and heart. The concentrations in other organs are relatively low.

Similar observations have been reported for rats (Friberg, 1956), sheep and foxes (Buzadshi, 1965) and hens (Ulfvarson, 1969). Platonow (1968) has reported that, after the intramuscular injection of methylmercury acetate and ethylmercury acetate, a considerable increase in the mercury concentration of the central nervous system occurred.

Table 6.5 Average Hg Content of the Blood (Erythrocytes), Plasma and Hair of Regular Fish-eaters in Sees Vänern, Sweden (Tejning, 1967)

Group	Size	Age	Sample	Hg Content (ng/g)	Consumption (kg/week)
1	14	49 ± 3.8	blood	37 ± 8.8	3.0 ± 0.2
1	14	49 ± 3.8	plasma	4.8 ± 1.0	3.0 ± 0.2
1	14	49 ± 3.8	hair	4716 ± 1076	3.0 ± 0.2
2	12	55 ± 4.6	blood	91 ± 15.0	3.0 ± 0.2
2	12	55 ± 4.6	plasma	13.4 ± 2.9	3.0 ± 0.2
2	12	55 ± 4.6	hair	11720 ± 2490	3.0 ± 0.2
3	10	53 ± 4.1	blood	59 ± 15.8	3.7 ± 0.6
3	10	53 ± 4.1	plasma	5.8 ± 1.0	3.7 ± 0.6
3	10	53 ± 4.1	hair	8225 ± 1627	3.7 ± 0.6
4	15	51 ± 3.8	blood	51 ± 6.6	3.0 ± 0.2
4	15	51 ± 3.8	plasma	6.4 ± 0.8	3.0 ± 0.2
4	15	51 ± 3.8	hair	7638 ± 1125	3.0 ± 0.2

Reith and Gerritsma (1955) estimated the daily absorption of mercury by human beings to be of the order of 0.020mg. Small amounts of this element are normally present in all of the organs of an individual. It seems that the kidneys contain the highest concentration of mercury (Forney, 1950; Stewart & Stolman, 1960). The normal mercury contents of the various organs of human beings are indicated in table 6.1.

Hg CONTENT OF BLOOD

Goldwater et al (1964) analysed 609 blood samples from 16 different countries (table 6.4). In 25% of these samples, the values ranged from 0.09 to 0.32g/l of blood, while the rest were almost mercury-free.

Analyses carried out in Sweden (Tejning, 1967) revealed a 5-fold to 6-fold increase in the mercury concentration in the hair and blood samples of regular fish-eaters at Sees Vänern. The results of analyses of blood, plasma, and hair samples are shown in table 6.5.

A 60 year old fisherman living in Sees Vänern had the highest mercury concentration: 260ng/g in the blood cells and 27600ng/g in the hair. He had eaten 3 to 4kg of fish every day (Ackefors, 1968).

Hg TRANSPORT IN BLOOD

Lundgren et al (1967) determined the distribution of various mercury compounds in blood. Intravenous injections of $^{203}Hg(NO_3)_2$ in rats resulted in concentrations of 40 to 75% of total mercury in erythrocytes and 75 to 89% in plasma from the albumen fraction. However, in vitro experiments indicated that mercury combined with the plasma.

Mercury forms salts with proteins, alkalies in the blood, and other compounds which are soluble in tissue fluids. The thiol of human blood plasma, mostly found as serum mercaptalbumin, contains one SH- group per protein molecule. Some 5 to 10% of the total SH- groups of plasma are found in small molecules. Mercury can combine with these compounds and be transported to other organs very easily.

In vitro and in vivo studies by Takeda et al (1968) have shown that ethylmercury chloride has a strong affinity for the inner side of the erythrocyte membrane. Sahagian (1967) has found that the transmural exchange of mercury takes place with other cations. Mercury strongly decreases the transport of Mn but increases the transport of Cd.

KIDNEYS AND LIVER

After the administration of oral doses of mercury compounds, the highest concentration in the human body is observed in the kidneys, followed by the liver and milk (Sollman & Schreiber, 1936). The mercury is confined to the proteins of the tissues of the kidney, where its pathological effect is exercised. Experiments conducted by Swensson and Ulfvarson (1968) have revealed that mercury is particularly accumulated in the kidneys, brain, and testes when methylmercury (II) hydroxide or mercury nitrate is injected into rats. The removal of this metal from the organs takes place very slowly.

The kidney is the organ where the highest concentration of mercury is found. The metal ion accumulates in the collecting tubule, in the distal and proximal tubules, and also in the Henley loop, but not in the glomerulus (Bergstrand et al, 1952). Mercury has also been found in the epithelial layers of Bowman's capsule as well as in the endothelial cells in the arterioles of the kidneys.

Table 6.6 Mercury Content of Human Stones

Sample	Number	Parameter	Value	Units
Gallstones	10	range	2.6 - 7.9	ppm
Gallstones	10	average	4.8	ppm
Urinary calculi	9	range	0.1 - 5.4	ppm
Urinary calculi	9	average	2.6	ppm
Blood	8	range	0.34 - 0.63	μg/ml
Blood	8	average	0.46	μg/ml
Bile	8	range	0.14 - 1.7	μg/ml
Bile	8	average	0.54	μg/ml
Urine	8	range	10 - 52	μg/l
Urine	8	average	25	μg/l

HUMAN MILK

Fujita and Tatabalake (1977) determined the total mercury level (3.6ppb) in the breast milk of human mothers from the Tokyo metropolitan area. The mean maternal and neonate blood mercury levels were 3.3 and 4.3ppm respectively. Mercury concentrations in neonatal hair and maternal blood were significantly correlated.

Juszkiewicz et al (1975) observed that the residues of mercury in the milk from 101 women ranged from 0.0011 to 0.0158ppm and were higher in the milk of women who were more than 30 years old. Some differences were also observed between the Hg content of milk from urban women, and from rural women employed in the chemical preservation of seed grain for agriculture. Human milk, mainly colostrum and transitional milk, collected in the Ljubljana area of Yugoslavia, were found to contain mercury levels ranging from 1.2 to 37.4 μg/kg (average: 11.8 $\pm$ 8.3) (dry weight basis;

Kosta et al, 1983).

HUMAN CALCULI

Ohta and Terai (1971) studied the mercury content of human stones obtained from people who had not been exposed to mercury. The content of the metal was determined photoelectrically using dithizone. The results are shown in table 6.6, together with the values for blood, bile, and urine.

HAIR

The effects, of diets having a high sea-fish content, upon the Hg^{2+} content of human hair has been determined (Establier, 1975). The mean mercury (Hg^{2+}) level of hair was 4.08ppm for a population with an average sea-fish consumption of 1 to 4 meals per week. In fisherman whose sea-fish consumption was greater than that of the average population, the mean Hg^{2+} content was 19.81ppm. However, none of the subjects exhibited symptoms of Hg^{2+} poisoning. When swordfish were eliminated from the diet of a child, the mercury content of its hair decreased from 17.5 to 3.5ppm. Suzuki et al (1972) pointed out that marine fish might be the major factor in the transfer of Hg from the environment to man. Mercury levels in the hair of some Japanese people living in Burma or Bangladesh changed with food availability. After residing in an inland area of Burma for 6 months, the levels of Hg in the hair of a Japanese person decreased from 4.28 to 1.72ng/g. The diet consisted mainly of locally mixed foods. In the hair of those residing in a coastal town in Bangladesh for 10 months, no change in the Hg level was found. Their diet consisted of locally caught marine fish and canned food from Japan. The biological half-life of Hg was calculated to be 33 days (less than the 70 day half-life of methylmercury in man). Several studies indicate that the hair normally reflects an individual's body content of mercury (Airey, 1983b). The hair usually contains 300 times more mercury than does the blood. This fact makes analysis easier (Phelps et al, 1980) and, in addition, hair does not require refrigerated storage facilities. It grows at the rate of about 1cm per month.

The Hg content of hair varies with differences in dietary habits (especially fish intake) rather than in living environments. The Hg concentrations in new-growth hair indicates blood Hg concentrations, or the body burden of mercury existing during the period of growth (Sexton et al, 1978; Kershaw et al, 1980). The hair Hg levels in the fishermen and crewmen of tuna fishing boats are higher than those in the general population. The hair of vegetarians might indicate environmental effects which are not related to fish in the diet.

Higher methylmercury levels were detected in the head and underarm hair of sericultural workers who had been constantly exposed to phenylmercuric acetate until the discontinuation of the use of this compound as a disinfectant in 1971. The underarm, crural, and pubic hair have relatively low Hg levels, compared to the head hair. But these, as well as the nails, can also serve as indices of Hg exposure (Yamanaka et al, 1974).

Table 6.7 Mean Mercury Concentrations in the Hair of Persons who Eat Fish 1 to 4 Times per Month

Country	Content (ppm)
Australia	2.5
Canada	1.2
China	0.9
West Germany	0.5
Hong Kong	3.0
Italy	1.5
Japan	3.9
Monaco	1.7
New Zealand	1.3
Papua New Guinea	1.8
South Africa	1.9
UK	1.6
USA	2.4

When 559 samples of human head hair, from people who ate fish 1 to 4 times each month and lived in 32 locations in 13 countries, were analysed, the mercury concentrations were as shown in table 6.7. The differences are due to eating habits and environment.

The mean hair concentrations were significantly different for the group whose members ate fish once or less per month (1.4ppm), once per fortnight (1.9ppm), once per week (2.5ppm), or once or more per day (11.6ppm).

The weighted mean mercury concentration in the hair of people living both in the northern and southern hemispheres, at latitudes greater than 40°, was significantly lower than that for other latitudes, even in industrialised countries. It is suggested that lower soil temperatures could retard mercury volatilisation and inhibit its movement into ecological cycles which involve man (Airey, 1983b).

A group of seamen, recruited from all over the USA to attend a naval academy, exhibited a large standard deviation about the mean hair concentration due to the widely differing lifestyles of its members (Gordus et al, 1974). After a year or two at the academy, where they followed comparatively similar lifestyles and ate the same food, the standard deviation about the mean decreased considerably. Other studies (Suzuki et al, 1972; Al-Shahristani & Al-Haddad, 1973; Giovanoli-Jazubezak et al, 1974) showed that, when people emigrated from Japan to Burma, from Canada to Iraq, or from Japan to Rochester, NY, they exhibited a significant decrease in the mercury content of their hair.

Ohno et al (1967) have observed a Hg concentration of 7.5 ± 4.8ppm in the hair of 17 farmers who had handled mercury fungicides, and a concentration of 9.8 ± 2.9ppm in the hair of dentists.

Ishizaki et al (1975) determined the mercury levels in the hair of new-born children over a period of one year. The levels were higher in male, than in female, subjects. The Hg levels remained constant for 6 months after birth and then tended to increase. The concentrations of Hg and Zn in human hair samples collected throughout Japan were determined by Takahashi et al (1976), using NAA. The mean value for Hg was 6.4ppm. Ohta (1978) has reported values of 3.7ppm and 1.3ppm for residents of Japan and France, respectively. The difference was attributed to the greater consumption of sea-food in Japan.

The value for hair samples from a dentist was 11.3ppm. A significant difference in the Zn content of hair was observed between males and females. Wang and Yeh (1974) surveyed the mercury content in the hair of Taipei citizens, where the average Hg content was 2.68ppm. The mercury level was higher (3.04ppm) in females than in males (2.36ppm). The contents appeared to be within normal limits and increased with age from 6 to 40 years.

In a survey of the mercury contents of hair from 1000 people living in locations throughout Idaho, the mean Hg levels for samples from males and females were 2.45

and 5.90ppm, respectively. The levels ranged from a low value of 0.12ppm in a male to a high value of 139ppm in a female (Benson & Grabica, 1972).

The total mercury contents in the hair of 848 subjects among the inhabitants of Shinnauyo city, Tokushima city, and Ube city were determined to be 3.26, 3.50, and 3.30ppm, respectively. No significant difference was found between these values and the Japanese average (Furuno et al, 1974).

Five varieties of canned fish, and hair specimens from 243 subjects in Port Moresby, were analysed for their organic and total mercury contents by Kyle and Ghani (1983). The mean hair concentration was 2.1ppm, with a range of 0.15 - 14.4ppm (standard deviation: 0.8ppm). The average percentage of organic mercury in the total content was 80% (table 6.8). The total mercury levels in hair were significantly higher for coastal populations than for non-coastal populations.

Table 6.8 Mercury Concentrations in the Hair of Coastal and Non-Coastal Populations

Mercury Fraction	Population	Sample Size	Parameter	Value (ppm)
Organic	coastal	121	mean $\pm$ SD	2.1 $\pm$ 1.7
Total	coastal	121	mean $\pm$ SD	2.6 $\pm$ 2.2
Organic	non-coastal	122	mean $\pm$ SD	1.3 $\pm$ 0.9
Total	non-coastal	122	mean $\pm$ SD	1.7 $\pm$ 1.1
Organic	coastal	121	range	0.3 - 11.6
Total	coastal	121	range	0.4 - 14.4
Organic	non-coastal	122	range	0.1 - 4.0
Total	non-coastal	122	range	0.2 - 5.2

On the basis of a study of 753 people in Japan, Kojima et al (1979) have found a relationship between the daily amount of methylmercury eaten in the form of fish and the mercury content of hair. That is:

$$\text{Mercury in hair (ppm)} = 150 \times \text{methylmercury in fish eaten (mg)} + 1.66$$

Airey (1983a) has compiled data on the mercury levels in hair of human populations throughout the world, and correlated them with the gross national product (GNP), thermal energy, and fish consumption. His studies show that mercury is being accumulated by people living in industrial nations situated in the northern hemisphere at latitudes above 22°N (Airey, 1982). The weighted mean concentrations of mercury in human head hair sampled from the southern hemisphere, tropical northern hemisphere (0 - 22°N), and above 22°N, are shown in table 6.9

Table 6.9 Mercury Content of Human Hair Samples

Latitude	Total Number of Samples	Weighted Mean (ppm)
Below 0°	1629	1.685
0 to 22°N	145	2.318
Above 22°N	5546	3.809

Recent work has suggested the possibility of minor defects occurring in babies born to mothers having hair mercury levels greater than 10ppm (Piotrowski & Inskip, 1981).

NAILS

The Hg levels in thumb-nails were determined by Kanabrocki (1977), using a thermal neutron activation technique. Samples were obtained from 14 subjects employed in the hospital dental service (a), from 19 employed in clinical laboratories (b), and from 16 hospitalised patients who were not employed in laboratory occupations (c). The average Hg levels were (a) dentists - 2.7ppm; technicians - 3.9ppm; assistants - 6.6ppm, (b) male technicians - 0.4ppm, female technicians - 1.9ppm, (c) patients - 0.3ppm.

ABSORPTION IN SMALL INTESTINE

Experiments by Sahagian et al (1966) indicate that the absorption of mercury is greater in the duodenum and ileum than in the jejunum of rats. The duodenum and ileum exhibit little difference in absorption ability. The absorption rate is lower than that of Zn but faster than that of Cd and Mn. The jejunum absorbs the lowest amount of mercury in comparison to the other elements of group IIB of the periodic table i.e., Zn and Cd.

Sahagian et al (1966) have also studied the influence of the elements, Zn, Mn, Cd, upon the absorption of mercury in the small intestines of rats. Mercury is not transported in the small intestine by any concentration gradient. The absorption of mercury is concurrent with the metals, Zn, Cd, and Mn. It has also been found that absorption of the metal through the mucosa is possible.

INTRACELLULAR DISTRIBUTION

Alkylmercury compounds are easily absorbed and evenly distributed in many tissues without decomposition. Swensson (1971) suggested that aryl and alkoxyalkyl derivatives were less easily absorbed, decomposed mainly to inorganic mercury, and accumulated principally in the kidneys and liver. Mercury is not uniformly distributed within the cell. Berlin and Ulberg (1963) have detected mercury deposition in the nucleus. Using chemical fractionation of rat brains, Hashino (1965) found very little mercury to be present. All of the activity of the radio-isotopes used was located in the protein.

Norseth (1967, 1968) separated the sub-cellular particles from the liver of rats after poisoning them with methoxyethylmercury, and found an accumulation of mercury in the lysosome- and microsome-rich fractions.

Wester (1964) studied the mercury concentration in the sub-cellular fractions of the heart tissue of bulls. He found high mercury concentrations in the michondrial fraction.

RATE OF EXCRETION

The excretion of mercury takes place through the kidneys and intestines, and via bile, stools, and milk. The excretion is rapid when the intake is large. After removal of the contaminating source, the excretion continues for six months or more.

Ulfvarson (1969) described the excretion by hens, of orally administered and intravenously injected mercury compounds, using a hyperbolic function:

$$1/x = Kt + 1/x_o$$

where x is the concentration in the body after t days, x_o is the original concentration, and K is a constant having a value of 10^{-5}[g(Hg)/g(body weight)]/day for $Hg(NO_3)_2$, methoxyethylmercury hydroxide, and phenylmercury hydroxide, and only 40% of this value for methylmercury hydroxide.

URINE

A concentration of 5mg/day in urine has been reported by many workers. The results of normal urine samples from 14 different countries are outlined in table 6.10.

Table 6.10 Normal Values of Mercury in Urine (Jacobs et al, 1964)

Country	Samples	Range (μg/l)	Number of Samples with Hg
Argentina	49	0 - 21.0	41
Chile	27	0 - 37.5	24
Czechoslovakia	20	0 - 10.5	17
Egypt	14	0 - 12.0	9
England	30	0 - 37.0	26
Finland	45	0 - 22.5	30
Israel	52	0 - 94.5	48
Italy	30	0 - 37.5	24
Japan	40	0 - 45.0	34
Netherlands	42	0 - 4.5	36
Poland	66	0 - 6.0	57
Sweden	30	0 - 73.5	24
Yugoslavia	45	0 - 7.5	40
USA	308	0 - 221.0	240
Overall	778	0 - 221.0	629 (80.8%)

The biological half-lives of mercury compounds in various mammals have been discussed earlier.

BIOLOGICAL HALF-LIFE

Alkylmercury compounds are stable and are slowly excreted, whereas inorganic mercury compounds, as well as phenyl and alkoxyalkylmercury compounds, have a half-life of 3 to 4 days in mammals. In rats, it is 15 days, while it is 23 to 27 days for poultry. Following the administration of ^{203}Hg-labelled compounds to rats, Yasumasa et al (1968) observed the following half-life values:

Phenylmercury acetate	22.0 days
Phenylmercury chloride	27.9 days
Ethylmercury chloride	15.3 to 48.8 days

The study of Iraqi mercury poisoning cases indicates that the biological half-life of methylmercury can vary from 35 to 120 days, with 72 days being the average.
The biological half-life of mercury, following the intraperitoneal injection of methylmercury dicyandiamide into rats, is 40 days (Berglund & Wretlind, 1969). Ekman et al (1968) studied the turnover of mercury in the human organism. The biological half-life of ^{203}Hg was found to be 70 to 74 days (1mg $CH_3^{203}HgNO_3$/week, orally administered). The effect of mercury varies from subject to subject, and this is not related to age, gender, or eating habits.

REFERENCES

- Ackefors, H. (1968) - International Council for the Exploration of the Sea C.M. 1968/E:3, Fisheries Improvement Committee
- Airey, D. (1982) - Science of the Total Environment, 25, 19-40
- Airey, D. (1983a) - CSIRO Australia, Marine laboratory, Microfiche reprograph, 4
- Airey, D. (1983b) - Science of the Total Environment, 31, 157-80
- Al-Shahristani, H. and Al-Haddad, I.K. (1973) - Nuclear Research Institute, Tuwaitha, Baghdad, Report No. PH-15, pp 1-12
- Battigelli, M.C. (1960) - Journal of Occupational Medicine, 2, 337-44 and 394-9
- Benson, W.W. and Grabica, J. (1972) - Pesticides Monitoring Journal, 6(2), 91-3
- Berg, K., Wanntorp, H., Erne, K., and Hanko, E. (1966) - Mercury Poisoning in Wild Life in Sweden, mimeographed report from the National Veterinary Institute, 1, Stockholm (in Swedish)
- Berg, K., Wanntorp, H., Erne, K., and Hanko, E. (1969) - Viltrevy, 6(4), 301
- Berglund, F. and Wretlind, A. (1969) - Nordisk Hygienisk Tidskrift, 50(2), 118-24
- Bergstrand, A., Friberg, L., and Odebland, B. (1952) - American Archives of Industrial Health, 17, 253

- Berlin, M. and Ulberg, S. (1963) - Archives of Environmental Health, 6(5), 602-16
- Brune, D., Nordberg, G., and Wester, P.O. (1980) - Science of the Total Environment, 16, 13-35
- Buzadshi, A. (1965) - Sb. Nauch. Tr., Donkoi Sel skokhoz. Inst., 2(1), 48-52
- Clarkson, T.W. (1977) - in Clinical Chemistry and Chemical Toxicology of Metals (edited by S.S.Brown), Elsevier, Amsterdam, 189-200
- Curley, A., Sedlak, V.A., Garling, E.F., Hawk, R.E., Barthel, W.F., Pierce, P.E., and Likosky, W.H. (1971) - Science, 172, 65-7
- Drake, H.J. (1979) - Mercury. Bureau of Mines Minerals Yearbook, Volume 1, US Government Printing Office, 189-902
- Ekman, L., Greitz, U., Persson, G., and Aberg, B. (1968) - Nordisk Medicin, 79, 450-6
- Ellis, R. and Fang, S. (1967) - Toxicology and Applied Pharmacology, 11(1), 104-13
- Engleson, G. and Herner, T. (1952) - Acta Paediatrica, 41, 289
- Establier, R. (1975) - Investigacion Pesquera, 39(2), 509-16 (Spanish)
- Flaschentrager, E. (1951) - Physiologische Chemie, I, 208
- Forbes, R., Cooper, A., and Mitchell, H. (1954) - Journal of Biological Chemistry, 209, 857-66
- Forney, R. (1950) - Federation Proceedings, 8, 292
- Friberg, L. (1956) - Acta Pharmacologia et Toxicologica, 12, 411-27
- Friberg, L. and Nordberg, G.F. (1972) - in Mercury in the Environment (Edited by L.Friberg and J.Vostal), Chemical Rubber Company Press, Ohio, 113-39
- Friberg, L. and Vostal, J. (1972) - in Mercury in the Environment (Edited by L.Friberg and J.Vostal), Chemical Rubber Company Press, Ohio, 183-6
- Fujita, M. and Tatabalake, E. (1977) - Bulletin of Environmental Contamination and Toxicology, 18(2), 205-9
- Furuno, J., Sugawara, N., and Naitou, Y. (1974) - Igaku To Seibutugaku, 86(6), 443-7 (Japan)
- Giovanoli-Jazubezak, T., Greenwood, M.R., Smith, J.C., and Clarkson, T.W. (1974) - Clinical Chemistry, 20, 222-9
- Glomski, C.A., Brody, H., and Pillay, S.K.K. (1971) - Nature, 232(5301), 200-1

- Goldwater, C.J., Kleinfield, M., and Berger, A.R. (1956) - Archives of Industrial Health, 13, 245-9

- Goldwater, L., Ladd, A., and Jakobs, M. (1964) - Archives of Environmental Health, 9(4), 735-41

- Gordus, A.A., Maher, C.C., and Bird, G.C. (1974) - in Proceedings of the First Annual NSF Trace Contaminants Conference (at Oak Ridge National Laboratory), August 1973, 463-87

- Griffith, G., Butt, E., and Walker, J. (1954) - Annals of Internal Medicine, 41, 501-9

- Haq, I.U. (1963) - British Medical Journal, 1, 1579

- Ishizaki, M. et al (1975) - Ibaraki-Ken, Eisei Kenkyusho Nempo, 13, 63 (Japan)

- ITRC - Industrial Toxicological Research Center (1981) - Food and Animals as Indicators of Air Pollution, Report 1976-1980, Lucknow, India

- Jacobs, M.B., Ladd, S.E., and Goldwater, L. (1964) - Archives of Environmental Health, 9(4), 454-63

- Joselow, M. and Goldwater, L. (1967) - Archives of Environmental Health, 15(2), 155-9

- Joselow, M. and Weinberg, S. (1967) - Archives of Environmental Health, 15(1), 64-6

- Joselow, M., Goldwater, L., Alvarez, A., and Herndon, J. (1968) - Archives of Environmental Health, 17, 39-43

- Juszkiewicz, T., Teodor, S., and Radomanski, T. (1975) - Poliski Tygodnik Lekarski, 30(9), 365-6 (Polish)

- Kanabrocki, E.L. (1977) - International Journal of Nuclear Medicine and Biology, 4(2), 128-32

- Karppannen, E., Henrikson, K., and Nurmi, E. (1968) - Annales Agriculturae Fenniae Supplements, 7(1), 24-9

- Kershaw, T.G., Dhahir, P.J., and Clarkson, T.W. (1980) - Archives of Environmental Health, 85, 28-35

- Kojima, K. and Araki, T., quoted by Y.Takizawa (1979) - in The Biogeochemistry of Mercury in the Environment, J.O.Nriagu (editor), Elsevier North Holland Biomedical Press, Amsterdam, 361

- Kosta, L., Byrne, A.R., and Dermelj, M. (1983) - Science of the Total Environment, 29, 261-8

- Kurland, L.T., Faro, S.N., and Siedler, H. (1960) - World Neurology, 1, 370
- Kyle, J.H. and Ghani, N. (1983) - Science of the Total Environment, 26, 157-62
- Lihnell, D. and Stenmark, A. (1967) - Vaxtskyddsnotiser, 31, 36 (Swedish)
- Lundgren, K., Swensson, A., and Ulfvarson, U. (1967) - Scandinavian Journal of Clinical Laboratory Investigations, 20(2), 164-6
- Mitra, S. and Ghosh Dastidar, P. (1976) - Science and Culture, 42(9), 443-51
- Mottet, N.K. (1973) - in Mercury in the Western Environment (Edited by D.R.Buhler), Proceedings of a Workshop, Portland, Oregon, 1971, Corvallis, Oregon, 1973, 311-9
- Nixon, G., Paxton, G., and Smith, H. (1967) - Journal of Dental Research, 44, 654
- Norseth, T. (1967) - Biochemical Pharmacology, 16, 1645-54
- Norseth, T. (1968) - Biochemical Pharmacology, 17, 27-68
- Ohno, T., Kishida, H., and Shimuzu, N. (1967) - Gifu Yakka Digaku Kijo, 17, 88-91
- Ohta, Y. (1978) - Analysis, 6, 145-7
- Ohta, N. and Terai, M. (1971) - Nippon Kagaku Zasshi, 92(6), 519-21
- Phelps, R.W., Clarkson, T.W., Kershaw, T.G., and Wheatley, B. (1980) - Archives of Environmental Health, 35, 161-8
- Piotrowski, J.K. and Inskip, M.J. (1981) - Health Effects of Methylmercury. A technical report. Monitoring and Assessment Research Centre, Chelsea College, London, 65
- Platonow, N. (1968) - Occupational Health Review, 20(1-2), 1-8
- Reith, J. and Gerritsma, K. (1955) - Voding, 16, 832-943
- Rossi, L.C., Clement, G.F., and Santaroni, G. (1976) - Archives of Environmental Health, 3, 160-5
- Sahagian, B., Harding-Barlow, I., and Perry, H. (1966) - Journal of Nutrition, 90, 259-67
- Sexton, D.J., Powell, K.E., Liddle, J., and Sinrek, A. (1978) - Archives of Environmental Health, 33, 186-91
- Skerfving, S. and Vostal, J. (1972) - in Mercury in the Environment (Edited by L.Friberg & J.Vostal), Chemical Rubber Company Press, Ohio 93-107
- Sollman, M. and Schreiber, N. (1936) - AMA Archives of Internal Medicine, 57, 46-62

- Stewart, C. and Stolman, A. (1960) - Toxicology, Volume 11, Academic Press, New York, 206-8

- Stock, A. (1940) - Biochimische Zeitschrift, 304, 73

- Suzuki, T., Matsubara-Khan, J., and Matsuda, A. (1972) - Bulletin of Environmental Contamination and Toxicology, 7(1), 26-32

- Swensson, A. and Ulfvarson, U. (1968) - Acta Pharmacologia et Toxicologica, 26(3), 259-72

- Takahashi, Y., Tsuji, K., and Okano, M. (1976) - Eisei Kagaku, 22(5), 296-302 (Japan)

- Takeda, Y., Kunugi, T., Hoshino, O., and Ukita, T. (1968) - Toxicology and Applied Pharmacology, 13(2), 156-64

- Tejning, S. (1967) - Rapport 6702 06, Yrkesmedicinkliniken via Lasarettet i Lund. (Mimeo)

- Ulfvarson, U. (1969) - Toxicology and Applied Pharmacology, 15(3), 525-31

- Wang, K. and Yeh, C. (1974) - Pei I Hsuch Pao, 6, 57-63 (Chinese)

- Warren, H.V., Horsky, S.J., and Gould, C.E. (1983) - Science of the Total Environment, 29, 163-9

- Watson, W.D. (1979) - in The Biogeochemistry of Mercury in the Environment (Edited by J.Nriagu), Elsevier North-Holland Biomedical Press, New York, 56

- Wester, P. (1964) - Aktiebolaget Atomergic AE-169, Stockholm, Sweden, 20

- WHO (1976) - Environmental Health Criteria, I - Mercury, World Health Organisation, Geneva

- Yamanaka, S. and Ueda, K. (1974) - Sangyo Igaku, 61(2), 132-3 (Japanese)

- Yashino, K. (1965) - Rinsho Shinkeigaku, 5(3), 130-44

- Yasumasa, K., Urakubo, G., and Hasegawa, A. (1968) - Eisei Kagaku, 14(2), 76-81

CHAPTER VII

MERCURY IN PLANTS, BIRDS, AND FISH

INTRODUCTION

Metals such as Cr, Pb, and Hg are more concentrated in the roots than in the leaves or stems of plants. Soil which is enriched in mercurial compounds due to fungicide spraying and industrial pollution may supply plants with soluble mercurials but, normally, plants grown in soil which contains low amounts of mercury do not accumulate it to a significant degree.

PLANTS GROWN IN A NORMAL ENVIRONMENT

Reports on the analysis of mercury in plants have shown that this metal is not accumulated to a great extent in the tissues of most plants grown in normal environments. Hawkes and Webb (1962) have reported that the concentration of mercury in vegetation is 30-300ppt, and Warren and Delavault (1969) have indicated a value of 10-60ppt. In a biogeochemical investigation conducted in Missouri, the US Geological Survey, (Shacklette, 1970) reported that the concentration of mercury in dry plant material was 500ppt. The study involved 96 native trees and shrubs of species such as ***Quercus stellata***, ***Q. alba***, ***Q. lyrata***, ***R. copallina***, ***Rhus glabra***, and ***Juniperus virginiana***. In an associated study of roadside contamination of vegetation and soils in Missouri, 33 red cedar samples were found to contain less than 500ppt of Hg, whereas the mercury content of dry samples of low soils in which these trees were rooted ranged from 40 to 650ppt. The samples selected were the terminal parts of the stems of deciduous trees and shrubs and the terminal branches, including leaves, of red cedar.

The normal values of mercury in vegetables and fruits are indicated in table 7.1 (Stock & Cucuel, 1934) and table 7.2.

Hg CONCENTRATION IN CEREALS

Smart and Hill (1968) analysed 26 rice samples from Argentina, Australia, China, Korea, Pakistan, Thailand, and the USA, and observed a mercury concentration of 0.005ppm in 17 samples. However, Japanese researchers (Suzuki, 1959; Tankawa, 1969), found a higher mercury concentration (0.12ppm).

Table 7.1 Analysed Values of Mercury in Food Products of Plant Origin

Product	Hg Content (ppm)
Cabbage	0.007
Lettuce	0.009
Beetroot	0.028
Horse-radish	0.044
Potato	0.001-0.004
Strawberry	0.007
Cherry	0.005
Apple	0.004
Pear	0.012
Peach	0.012
Rye flour	0.036
Wheat flour	0.020
Olive oil	0.012
Coconut oil	0.062
Palm oil	0.086
White wine	0.004
Vinegar	0.014
Beer	0.015
Algae	0.018-0.037

Research carried out by Furutani and Osajima (1965) indicated that the mercury concentration in rice was 0.2ppm, and that polishing of the rice removed about 30% of the mercury. A further reduction in the mercury concentration was observed after cooking. The fact that the concentration is higher in raw rice than in polished rice indicates the presence of mercury in rice husks.

Nadkarni and Ehman (1971) analysed US commercial wheat flour samples using neutron activation techniques. The results obtained for mercury were in good agreement with the recommended values given by the IAEA, that is, 0.5ppm.

Table 7.2 Analysed Values of Mercury in Food Products of Plant Origin

Product	Hg Content (ppm)	Reference
Apple	0.004	Pickard & Martin, 1960
Apple	0.004	Jacobs & Goldwater, 1961
Cabbage	0.159	Ruzicka & Lamm, 1969
Chestnut	0.02-0.10	Furutani & Osajima, 1967
Corn	0.044	Ruzicka & Lamm, 1969
Polished rice	0.04-0.12	Tankawa, 1969
Potato	0.0025	Legatowa et al, 1970
Rice	0.2	Furutani & Osajima, 1965
Tomato	0.010	Pickard & Martin, 1960
Wheat flour	0.01-0.10	Furutani & Osajima, 1967

Hg CONCENTRATION IN APPLES

Gutenmann and Lisk (1960) found a mercury concentration of 0.18ppm in apples which had been treated with mercury fungicides. Jacobs and Goldwater (1961) observed that apples which had been treated with mercury fungicides during the early stages of growth contained a greater concentration of mercury than did those which were treated in the later stages. The fungicide-treated apples had a concentration which was 0.02ppm higher than that in the untreated ones.

On average, 0.13ppm of Hg was found in the skin or peel of apples, 0.04ppm was found in the pulp, and 0.2ppm was found in the concentrated juice (Martin & Pickard, 1957). Chichoszowa et al (1969) determined the concentration of mercury in apple trees. The bud and fruit stages of the tree (75 - 120 days before harvesting) contained 0.052 to 0.083mg/kg. Small apples which had been sprayed with mercury compounds contained

0.0225 to 0.038mg/kg, but no mercury was detected after 60 - 90 days.
Chan Chen-Chang et al (1977) analysed orange samples by using neutron activation techniques, and all of the observed mercury levels were found to be below the permissible levels established in the US and other countries.

Table 7.3 Range of Mercury Concentrations in Vegetables, and Standard Values (Fresh Weights) which are Accepted in the FRG and Netherlands

Country	Range (ppm)	Reference
FRG	0.2 - 33	Bundesgesundheitsblatt, 1979
Netherlands	1 - 25	Staarink & Hakkenbrak, 1982
Standard	20 - 30	LAC Annual Report, 1980

MERCURY IN EDIBLES

The presence of mercury in edibles such as plants, vegetables, fish, fats, and oils, has been detected by various workers in recent years. Shitahara and Yasumasa (1976) determined the total mercury content of food by subsequent reduction vaporization. The fish contained up to 0.1ppm, but none exceeded the regulation level. High levels of mercury were detected in rice, soybeans, green peas, and mushrooms, but they were low in green vegetables, fruit, and potatoes. The mercury contents of pork, chicken, and beef were 0.021, 0.012, and 0.005ppm, respectively. Dairy products contained the least mercury; with levels as low as 0.001ppm. The total intake of mercury calculated on the basis of these data was 17.5μg/person-day; half of which came from fish and marine products.

The mercury contents of 16 edible mushroom species from various geographical areas were determined using flameless atomic absorption spectroscopy. The mercury contents of the species, ***Boletaceae***, ***Lycoperdacea***, ***Agaricaceae***, etc, exhibited constant levels of more than 1000ppb. However, in ***Canthatellacea*** it was less than 100ppb. Apart from other factors, the mercury concentration was essentially dependent upon the protein content (Aichberger, 1977).

Woidich and Pfannhauser (1975) estimated the mercury content of edible fungi to be between 0.001 and 3.22ppm. They observed that the mercury concentration was relatively high in ***Boletus*** sp and ***Agaricus campestris***, and low in ***Russula Agaricus biosporus*** and ***Cantharellus cibarius***. The possibility of mercury uptake, and potential accumulation mechanisms, were discussed by them.

A comparison of the metal contents of ***Aster*** and ***Salicornia***, from W.Scheld in the south-west of the Netherlands, with those found in vegetables in Germany (FRG) and the Netherlands showed that these natural vegetables contained accumulations of 0.3 to 0.4 times the upper limit of 30ppb accepted for residues left in vegetables (table 7.3). In ***Aster*** and ***Salicornia***, the values range from 20-30 to 50-70ppb on a dry weight (= 10 x fresh weight) basis (Beeftink & Nieuwenhuize, 1982).

The range of mercury contamination in the soils and plants of Alexandria (Egypt), due to industry (paper factories, chlor-alkali plants, electricity generating stations) and agriculture, was studied by El-Sokkary (1982). He found that highly polluted soils contained higher levels of mercury (495ppb) than did unpolluted areas (35ppb). The topsoil was highly enriched with mercury as compared to the subsoil. This was the result of mercury being disseminated from industrial discharges to the atmosphere and the soil. Vegetables (lettuce, radish, parsley, carrots, cabbage, peppers, etc) grown in soils near to industry contained higher levels of mercury (up to 362ppb). The amounts of mercury in plant species which grew in winter were higher than those in plants which grew in summer.

The published concentrations of trace metals in edibles and fats (Thomas, 1976) vary considerably. The values determined for Pb and mercury in crude oils and fats are quite often above the limits suggested by German legislation. As a result of refining, which is normally carried out on fats and oils, the mercury limit is sometimes exceeded. The neutralisation step is primarily responsible for a reduction in Pb content. Thomas has given values for As, Pb, Hg, and Cd in some materials, including soybean and coconut oil, at various stages of refining.

Pallotti et al (1976) analysed both edible and poisonous wild mushrooms: 148 samples from 64 species of 23 genera. The highest value (dry weight basis) was 11.40ppm, and the lowest value was 0.04ppm. ***Psalliota sivicola*** and ***Psalliota xanthoderma*** had the highest average levels; the lowest were found in ***Cantharellus cibarius***, ***Pholiota squarrosa***, ***Polyporus sulphureus***, and ***Polyporus giganteus***. Even higher values than these were found in mushrooms which were growing on mercury-rich soil (e.g., in the vicinity of cinnabar mines), and the analysis of mushrooms growing in various soils showed that they tended to concentrate mercury from the soil. Analysis of common samples of ***Boletus edulis*** and

Psalliota hortensis revealed average mercury levels of 1.25 and 0.61ppm (dry weight), respectively. Mercury in mushrooms will be discussed further.

Table 7.4 Mercury Levels in Some Plant Species
(From Geological Survey Professional Paper 713, p35)

Species	Part	Samples	Content*
Alder (***Alnus crispa*** ss ***Crispa***)	stem	1	1000
Black spruce (***Pisea mariana***)	stem, leaf	4	1000 - 1500
Dwarf birch (***Betula nana***)	stem	6	500 - 1000
Labrador tea (***Ledum palustre***, ss ***decumbem***)	stem	7	1000 - 3500
Spirala (***Spirala beauverdiana***)	stem, leaf	1	3000
White birch (***Betula papyrifera***, ss ***humilis***)	stem	4	500 - 2000

*ppt - dry plant

Beides and Higgons (1957) have determined the mercury concentration in tomatoes grown in greenhouses, and treated with mercury aerosols (phenylmercury chloride). Even 34 days after the aerosol treatment, 90% of the mercury was retained in the fruit pulp. The mercury concentration rose to 0.5ppm in harvested crops. Thus, mercury appears in the fruit market and is a potential health hazard for the consumer.

Hamilton and Ruthven (1967) analysed potato seeds which had been treated with organomercuric disinfectants for export purposes. The potatoes grown from the seed contained very small amounts of mercury and were indistinguishable from those grown from untreated seed.

PLANTS GROWN IN AN ENVIRONMENT CONTAINING ABNORMAL AMOUNTS OF MERCURY

Patton, in a survey of mercury and other elements in plants growing above cinnabar veins at Red Devil on the Kuskowim River (Lower Yukon River district, Alaska), reported the results given in table 7.4.

It should be noted that those trees whose roots penetrated through the loose mantle and came into contact with a cinnabar vein invariably contained large amounts of mercury in their branches, whereas neighbouring trees which were not in direct contact with the vein contained lower amounts of mercury.

Rankama and Sahama (1950) stated that droplets of metallic mercury had been found in the seed capsules of ***Holosteum umbellatum*** (chickweed of the family ***Caryophyllaceae***), growing on some mercury-rich soils. Marine algae may concentrate mercury, and some species are found which contain more than a hundred times as much mercury as does sea-water. In exceptional cases, mercury is concentrated as native mercury in some land plants. Vegetable fats are relatively rich in mercury.

The occurrence of drops of metallic mercury under the moss cover of the forest floor near to hydrothermal mercury deposits in the Rhine Palatinate was reported by Goldschmidt (1954). However, a search, organised by the US Geological Survey, for evidence of the mercury contamination of plants growing adjacent to a mercury smelter at Red Devil revealed none.

Mercury compounds have been widely used in agriculture as fungicides and appear to be quite harmless to the plant system. Some floriculturists believe that volatilised elemental mercury is toxic to certain crops, especially roses.

Shacklette (1965) reported the evidence for mercury poisoning, in the Red Devil area, in the following terms, "Presumably, the soil in the vicinity of the mine, mill, and smelter has been contaminated as a result of several years of operation of these installations. However, both bryophytes (mosses and liverworts) and vascular plants appeared to be remarkably unaffected. Mosses common to the region grew in a cinnabar mill and smelter drainage stream in which metallic mercury could be seen, and plants grew on a mountain tundra slope immediately adjacent to, and on a level with, the mercury. Smelter exhaust stacks appeared undamaged. No undisturbed outcrops of cinnabar that bryophytes could have colonised were found, but cinnabar was found in placer deposits and in rock used to surface a road, as well as around the mine shafts, and it did not appear to have had any effect on the mosses growing near it. We exposed some cinnabar outcrops by digging and found tree and shrub roots that were in contact with the mineral. Branches of the plants having root contact contained anomalous amounts of mercury yet the plants showed no toxicity symptoms".

Plants treated with mercury compounds may have high mercury concentrations in certain parts but it is not known whether the element was absorbed and translocated by plant tissues to these parts, or only occurred as a superficial residue.

Novick (1969) has provided evidence that plants are capable of absorbing mercury and translocating it through their tissues. Thus, mercury fungicides applied to leaves may be translocated to the fruit and other parts.

According to Harris et al (1970), organomercurial fungicides (0.1ppb) reduced the rate of photosynthesis and growth of plants, e.g. ***Nitzschia delicatissima*** and various ***Merismopedia***, ***Navicula***, ***Crucigenia***, ***Staurastrum***, and ***Ankistrodesmus*** species. This conclusion was prompted by laboratory tests. The compounds tested were AcoHgPh, MeHgNHC - (NH) NHCN, Ph_2Hg, and N-methylmercury - 1,2,4,6-tetrahydro 3,6-methano 3,4,5,6,7,7-hexachlorophthalimide (I). The highest toxicity to phytoplankton was shown by AcoHgPh and I, and the lowest by Ph_2Hg.

Bol'shakov et al (1970) analysed mercury from Wormwood and Yarrowtree in the region of mercury ore deposits containing 0.1 to 0.2ppm, and occasionally up to 0.5 - 1ppm. The mercury levels in the grass (10 species) and trees (3 species) reached several ppm in the factory area. The mercury content is distributed non-uniformly in the individual organs of the plant. The highest concentration of mercury was observed in the leaves and flowers (up to 20ppm), while the lowest was in the roots (0.4ppm).

Rauter (1975) determined the natural mercury contents of various wild mushrooms from various geological districts of Austria. The collected mushrooms contained about 10 times more mercury than normal and, occasionally, the values reached 100ppm (dry weight basis). Thus, commonly occurring ***Lycoperdon gemmatum***, ***Mycenapura***, and all ***Collibiaceae*** may be suitable indicators in the study of mercury pollution.

Loughton and Richard (1974) analysed mushroom samples, collected from Ontario (Canada), in which the mercury levels ranged from 0.05 to 0.33ppm. The mercury was present in an inorganic form.

An analysis of materials commonly used in the mushroom-growing industry revealed possible sources of contamination. Only horse manure and pesticides were considered to be likely to account for the exceptionally high residues. Horse manure is the most important source of possible contamination. The total mercury residues varied from 0.049 to 0.42ppm in dried manure and led to the production of composts with levels ranging from 0.074 to 0.61ppm.

MERCURY IN MARINE PLANTS

Marine phytoplankton organisms, such as unicellular algae, pick up mercury mainly by adsorption at their surfaces. Mercuric ion concentrations in water, due to the ultimate discharge of mercurial fungicides from the land, cause a considerable reduction in the photosynthetic capability of phytoplankton. Just 0.05ppm of Hg ions in water causes a

50% reduction in photosynthesis in the giant kelp, ***Macrocyctis pyrifera*** . A concentration of 500ppb causes a 15% decrease in photosynthesis in 1 day, and complete inactivation in 4 days.

Ukeles (1962) reported that 0.6ppb of ethylmercury phosphate was the threshold concentration for inhibition of the growth of marine phytoplankton, and that 60ppb was lethal to all marine species. Burrows and Combs (1958) concluded that ethylmercury phosphate was an effective algicide at a concentration of 100ppb. On the other hand, Helper and Olofsson (1975) reported that the threshold lethal concentration of mercury salts ranged from 900 to 60000ppb.

Sivic et al (1974) determined the mercury contents of the carbohydrate, lipid, and protein fractions of marine algae collected at sites suffering various levels of pollution. Mercury was found in the carbohydrate fraction, as well as in the proteins and lipids. This is significant because of the widespread use of algal carbohydrates in the food and pharmaceutical industries. Mercury concentration by algae also influences the path and level of this element in the food chain. Zingmark et al (1975) also studied the effect of an accumulation of mercury upon the phytoplankton.

When the organisms were shocked with warm or cold water it had a synergistic effect on production. When the phytoplankton were transferred from cold to hot water, production increased in the absence of mercury, whereas phytoplankton exposed to mercury were inhibited. Studies were also conducted in order to determine the effects of $HgCl_2$ upon the photosynthetic rate of phytoplankton and upon the growth and reproductive rates of selected cultural species of phytoplankton, as influenced by the concentration of mercury in the medium, the concentration in the cells, the stage of the growth cycle at which mercury exposure occurred, and the duration of exposure. The organisms used were the dinoflagellate, ***Amphidinium carterae*** , and the diatom, ***Skeletonema costatum*** . The effects of mercury and temperature interactions upon the growth of phytoplankton were studied using 3 species of algae. It was found that ***Synura petersenu*** was severely inhibited by mercury in all of the experiments. When ***Chlamydomonas*** was subjected to mercury at 30°C, growth was severely inhibited. The growth of ***Nitzschia*** was enhanced at 15°C when temperature shock was followed by the addition of mercury (Zingmark et al, 1975).

Corral and Masso (1975) found that the levels of mercury in plankton taken from the Arosa river in May, June, September, and October, 1974 ranged from 0.50 to 16.80ppm. Three of the 33 samples collected contained more than 10ppm, 16 samples contained 1 to 3ppm, and the remainder contained less than 1ppm.

BIRDS

Of the mammals, only small rodents can act as a significant link in the food chain which transfers methylmercury to predators. Since the mercury contents of small rodents have been rather low, it is understood that birds of prey which catch only rodents have lower body burdens of mercury than do those which catch only, or mostly, seed-eating birds.

Karppannen et al (1970) suggested the prohibition of the use of alkylmercury compounds for seed dressing. Gallinaceous and aquatic game birds which feed on aquatic animals have a markedly increased content of mercury as compared with terrestrial birds. However, the limited use of these game birds as food implies that the elevated mercury content is not an imminent threat to man.

According to Salei (1970), seeds treated with Granosan and used as food for hens, chicks, and swine led to the accumulation of mercury residues in their organs. The highest level was found in the liver, brain, and kidneys.

Baun et al (1977) analysed breast muscle from 591 band-tailed pigeons (***Columba fasciata***) collected in Colorado. Differences within gender and age classes were not significant. The percentage of pigeons with mercury levels greater than 0.50ppm was highest in 1969 and 1970 (34 and 29%, respectively), but decreased significantly in 1971 and remained low until 1974.

Using neutron activation techniques, Huckabee et al (1972) tested 148 samples of pheasant tissue for mercury contamination. Of these, 75% contained detectable mercury levels, 35% had mercury levels greater than 0.5ppm, and 2.7% had contents which exceeded 5.0ppm.

The reports of Dale (1970) indicate that pheasant tissue collected in 1968 at five locations in California (USA) had mercury levels greater than 0.15ppm. Griffith (1971) reported, on the basis of laboratory findings, that virtually all of the mercury in Californian pheasants was present in the form of methylmercury. The Alberta International Committee on Pesticides (1970) reported that pheasants and Hungarian partridges, collected from southern Alberta in 1969, contained an average concentration in the flesh of 0.45ppm of Hg, with a range of 0.24 to 0.79ppm.

Industrial wastes discharged into oceans lead to the accumulation of various toxic elements in marine organisms and marine food chains (Abdulla et al, 1972; Eisler et al,

1972; Dale et al, 1973, Peden et al, 1973; Koeman et al, 1975) and these ultimately become concentrated in the human system. Therefore, the determination of heavy metal residues in marine carnivore tissues, often pelagic and coastal fish-eating birds, is used to monitor the impact of toxic industrial waste upon marine food chains and to explain the decline of various avian populations (Connors et al, 1975; Fimreite, 1974; Fimreite et al, 1974; Gilbertson, 1974a,b).

Table 7.5 A Comparison of the Mean Cd, Hg, and Se Concentrations in Sooty Tern Eggs and Tissue from Dry Tortugas (Florida) and Lisianski Island (Hawaii), and the Mean Pb Concentrations in Eggs and Tissue from Lisianski Island

Material	Element	Location	Content (ppm - Wet Weight)
Eggs	Cd	Dry Tortugas	2.25 ± 0.09
Eggs	Hg	Dry Tortugas	7.93 ± 0.57
Eggs	Se	Dry Tortugas	3.78 ± 0.23
Eggs	Pb	Lisianski Island	0.38 ± 0.03
Eggs	Cd	Lisianski Island	75.04 ± 10.75
Eggs	Hg	Lisianski Island	0.15 ± 0.05
Eggs	Se	Lisianski Island	65.60 ± 5.38
Brain	Cd	Dry Tortugas	3.09 ± 0.06
Brain	Hg	Dry Tortugas	0.85 ± 0.05
Brain	Se	Dry Tortugas	4.73 ± 0.27
Brain	Pb	Lisianski Island	0.49 ± 0.09
Brain	Cd	Lisianski Island	0.16 ± 0.04
Brain	Hg	Lisianski Island	0.13 ± 0.18
Brain	Se	Lisianski Island	6.94 ± 1.78
Feathers	Cd	Dry Tortugas	4.25 ± 0.38
Feathers	Hg	Dry Tortugas	5.44 ± 0.92
Feathers	Se	Dry Tortugas	18.67 ± 0.89
Feathers	Pb	Lisianski Island	0.03 ± 0.02
Feathers	Cd	Lisianski Island	0.02 ± 0.01
Feathers	Hg	Lisianski Island	0.05 ± 0.01

continued

Table 7.5 (continued) A Comparison of the Mean Cd, Hg, and Se Concentrations in Sooty Tern Eggs and Tissue from Dry Tortugas (Florida) and Lisianski Island (Hawaii), and the Mean Pb Concentrations in Eggs and Tissue from Lisianski Island

Material	Element	Location	Content (ppm - Wet Weight)
Feathers	Se	Lisianski Island	0.77 ± 0.07
Kidneys	Cd	Dry Tortugas	23.50 ± 1.09
Kidneys	Hg	Dry Tortugas	1.64 ± 0.17
Kidneys	Se	Dry Tortugas	41.23 ± 3.51
Kidneys	Pb	Lisianski Island	1.73 ± 0.20
Kidneys	Cd	Lisianski Island	16.65 ± 1.79
Kidneys	Hg	Lisianski Island	1.36 ± 0.19
Kidneys	Se	Lisianski Island	10.96 ± 0.86
Liver	Cd	Dry Tortugas	3.79 ± 0.36
Liver	Hg	Dry Tortugas	1.52 ± 0.09
Liver	Se	Dry Tortugas	22.70 ± 1.24
Liver	Pb	Lisianski Island	0.03 ± 0.01
Liver	Cd	Lisianski Island	2.03 ± 0.19
Liver	Hg	Lisianski Island	1.59 ± 0.79
Liver	Se	Lisianski Island	18.08 ± 1.58

Ohlendorf et al (1978) reviewed the exposure of marine birds to environmental pollutants. The often high heavy metal residues in the tissues of birds which use coastal waters are generally accepted to be valid indicators of the toxic industrial waste load (Anderlini et al, 1972; Gilbertson, 1974a,b; Westermark et al, 1975).

Using neutron activation techniques, Hg, together with Cd and Se, were found (Stoneburner & Harrison 1981) in sooty tern tissue from the Gulf of Mexico (Dry Tortugas, Florida) and from the north central Pacific Ocean (Lisianski Island, Hawaii). The eggs collected from the Dry Tortugas were found to contain the highest mean concentration of mercury (7.93 ± 0.57ppm) (table 7.5).

Vermeer (1971) reported the results of a survey of mercury residues in composite egg samples from aquatic bird species, collected from various parts of the USA and Canada. He noted mercury residues of 0.1 to 0.4ppm for California gulls, 0.1 to 0.5ppm for ring-billed gulls, 0.1 to 0.3ppm for herring gulls, and 0.03 to 0.05ppm for Canada geese.

FISH

Fish assimilate mercury directly from the water as well as from their food. Comparison of the mercury concentrations in predatory fish and food fish showed that the former could absorb about 20% of the mercury present in the food fish (Jernelov, 1971). Also, Johnels et al (1967) indicated that the concentration factor from water to pike is of the order of 3000, while Bache et al (1971) observed a linear relationship between the age of the fish and the mercury content in trout from Cayaga Lake in Ithaca, New York. Moreover, Johnels and Westermark (1969) found a linear relationship between the age or weight of pike and the mercury content in the axial musculature or other organs. At low mercury levels (below 0.2ppm) in fish, very little increase in mercury content occurred with increasing weight. Similarly, at extremely high mercury levels, no relationship between age or gender and the mercury level existed. This was interpreted by Wallace et al (1971) to indicate that there is a threshold level of mercury in the environment, above which the rate of elimination of mercury from the muscle tissue of the fish is less rapid than its incorporation. Thus, accumulation occurs. It is apparent that fish have adapted to a mercury concentration of less than 0.2ppm (Lofroth, 1969). The relationship between size and mercury content in fish has necessitated the specification of a 'standard pike' as being a fish with a weight lying between 800 and 1600g. When comparing residues of mercury in fish samples for the purpose of restricting fishing, Sweden adjusted the level of mercury measured in pike so as to fit a fish weighing 1kg (Study Group on Mercury Hazards, 1970). According to Levitan et al (1974), some individuals of each of three carnivore fish species (***Sphyraena***, ***Chrysotaenia***, and ***Epinephulus guaza***) from Halifa and Akko bays contained more than 0.05ppm of mercury, with MeHg accounting for 70 to 100% of the total mercury content. The total content increased directly with increasing size (age) of the fish. All specimens of herbivorous fish (***Mugil capito***, ***M. Cephalus***, and ***Cyprinus carpio***) from these bays, and fresh water fish, contained less than 0.1ppm and usually less than 0.03ppm as total mercury content. No size correlation was observed in herbivorous fish. The total mercury content of water samples from both bays was 4 to 8 times higher than that of ponds, and was similar to that (0.1ppb) of the open sea. Lofroth (1969) observed that the mercury in fresh-water fish, and in Baltic and Atlantic fish is entirely in the form of methylmercury. Berglund et al (1971) have reported that the mercury in Swedish fish is almost entirely present as methylmercury. Noren and Westöö (1970), Johnels and Westermark (1969), and Westöö (1969) have all substantiated the fact that the mercury present in fish is entirely in the form of methylmercury. This is quite logical in view of the fact that all forms of mercury introduced into the aquatic environment can be converted into methylmercury (chapter IV).

Table 7.6 Mercury Concentrations in Canned Fish

Species	Number of Cans	Mercury Type	Parameter	Value (mg/kg*)
Tuna	38	organic	mean ± SD	0.35 ± 0.17
Tuna	38	organic	range	0.12 - 0.78
Tuna	38	total	mean ± SD	0.45 ± 0.22
Tuna	38	total	range	0.13 - 1.01
Salmon	10	organic	mean ± SD	0.16 ± 0.04
Salmon	10	organic	range	0.09 - 0.21
Salmon	10	total	mean ± SD	0.21 ± 0.05
Salmon	10	total	range	0.13 - 0.31
Mackerel	67	organic	mean ± SD	0.11 ± 0.07
Mackerel	67	organic	range	0.02 - 0.33
Mackerel	67	total	mean ± SD	0.17 ± 0.09
Mackerel	67	total	range	0.05 - 0.51
Sardine	5	organic	mean ± SD	0.07 ± 0.04
Sardine	5	organic	range	0.04 - 0.14
Sardine	5	total	mean ± SD	0.08 ± 0.04
Sardine	5	total	range	0.05 - 0.14
Pilchard	5	organic	mean ± SD	0.04 ± 0.02
Pilchard	5	organic	range	0.02 - 0.05
Pilchard	5	total	mean ± SD	0.04 ± 0.02
Pilchard	5	total	range	0.02 - 0.07

* wet weight

a) CANNED FISH

A British working party has published a summary report of an extensive survey of foodstuffs, including fish, with regard to mercury and other heavy metals (Ministry of Agriculture, 1971). Five varieties of canned fish from Port Moresby were analysed by Kyle and Ghani (1983). The mean concentrations of total mercury in the canned fish were as in table 7.6.

The mean mercury concentration in human hair in the region of this port is 2.1ppm. The total mercury levels in hair were found to be higher for more fish-eating coastal populations than for non-coastal populations.

b) PIKE

The concentration of mercury in samples of freeze-dried muscle from three different pikes was analysed by Gustavsson and Golimowski (1981) and were found to range from 4.3 to 22.7ppm.

Phillips and Gregory (1979) observed that the efficiency of methylmercury uptake by pike from food ranged from 58 to 89% in studies where the metal was added to synthetic foods and tissue homogenates, or fed to prey for short periods of time. However, when the pike obtained their methylmercury from contaminated prey throughout their life, the assimilation efficiencies were 20% or less. Longer exposure to the metal resulted in proportionately greater quantities of mercury occurring in slowly exchanging and more strongly bound pools of mercury in the organisms (Luoma, 1977).

Pike exposed to methylmercury for 70 to 90 days had the highest concentration in their kidneys, followed by the liver, spleen, stomach, heart, gills, brain, fins, gonads, muscles, scales, eyes, and bones. Salmon which were encaged in a river for two months exhibited an increase in the mercury content of the blood corpuscles, followed by the liver. The content in the muscle tissue increased slowly (Hasselrot, 1968). Jernelov (1970) stated that the ratio of the mercury content of the liver to that of the body muscle varied from 0.1 to 50. A liver/muscle mercury ratio which is greater than unity will indicate an increase in the content of fish muscle, while a ratio which is less than unity indicates that the content in the fish muscle will decrease or at least remain constant.

The uptake of methylmercury by pike is lower in brackish water than in fresh water.

However, cod concentrate methylmercury faster in salt water than in brackish water (Hannerz, 1968). Also, Larsson (1970) has observed that cod accumulate higher levels of mercury in sea-water than in brackish water. This is presumably because cod in 100% sea-water swallow more water than do cod in brackish water.

Table 7.7 Mercury in Various Varieties of Fish from the Colaba (Bombay) and Thana Markets

Market	Tissue	Hg Content (ng/g*)
	Karli (***Chiracentraus dorab***)	
Colaba	muscle	91 ± 4
Colaba	bone	177 ± 4
Thana	muscle	111 ± 4
Thana	bone	285 ± 8
Thana	brain	86 ± 2
	Mandeli (***Coilia dussumieri***)	
Colaba	muscle	60 ± 3
Colaba	bone	132 ± 4
Thana	muscle	70 ± 3
Thana	bone	146 ± 4
Thana	brain	41 ± 2
	Surmai (***Cybium guttatum***)	
Colaba	muscle	128 ± 6
Colaba	bone	304 ± 9
Thana	muscle	138 ± 4
Thana	bone	318 ± 10
Thana	brain	93 ± 3
	Rawas (***Eleutheron tetradactylus***)	
Colaba	muscle	126 ± 6

continued

Table 7.7 (continued) Mercury in Various Varieties of Fish from the Colaba (Bombay) and Thana Markets

Market	Tissue	Hg Content (ng/g*)
Colaba	bone	348 ± 16
Thana	muscle	187 ± 8
Thana	bone	329 ± 8
Thana	brain	107 ± 5
	Bombay Duck (***Harpodan nehereus***)	
Colaba	muscle	35 ± 1
Colaba	bone	112 ± 2
Thana	muscle	48 ± 2
Thana	bone	92 ± 2
Thana	brain	23 ± 1
	Bhingi, Indian Shad (***Hilsa ilisha***)	
Colaba	muscle	361 ± 10
Colaba	bone	493 ± 7
Thana	muscle	388 ± 11
Thana	bone	510 ± 18
Thana	brain	78 ± 3
	Red prawn (***Metapaenpios affinis***)	
Colaba	muscle	161 ± 6
Thana	muscle	180 ± 6
	Boi (***Mugil dussumieri***)	
Colaba	muscle	57 ± 2
Colaba	bone	143 ± 5
Thana	muscle	79 ± 2
Thana	bone	177 ± 4

continued

Table 7.7 (continued) Mercury in Various Varieties of Fish from the Colaba (Bombay) and Thana Markets

Market	Tissue	Hg Content (ng/g*)
Thana	brain	36 ± 1
	Pomphret (***Pampus argentius***)	
Colaba	muscle	92 ± 4
Colaba	bone	189 ± 7
Thana	muscle	124 ± 6
Thana	bone	205 ± 4
Thana	brain	73 ± 3
	Lobster (***Panalaris polyphagus***)	
Colaba	muscle	177 ± 8
Thana	muscle	207 ± 7
Thana	brain	58 ± 2
	Halwa (***Parastromateus niger***)	
Colaba	muscle	142 ± 7
Colaba	bone	312 ± 13
Thana	muscle	170 ± 4
Thana	bone	386 ± 9
Thana	brain	81 ± 4
	Salmon (***Polynemus tetradactylus***)	
Colaba	muscle	113 ± 4
Colaba	bone	297 ± 9
Thana	muscle	109 ± 4
Thana	bone	254 ± 3
Thana	brain	72 ± 3

continued

Table 7.7 (continued) Mercury in Various Varieties of Fish from the Colaba (Bombay) and Thana Markets

Market	Tissue	Hg Content (ng/g*)
	Sardine, Tarali (***Sardinella***)	
Colaba	muscle	77 ± 3
Colaba	bone	121 ± 2
Thana	muscle	66 ± 2
Thana	bone	133 ± 4
Thana	brain	40 ± 1
	Jew-fish, Dhoma (***Sciaena dicanthus***)	
Colaba	muscle	56 ± 2
Colaba	bone	137 ± 6
Thana	muscle	67 ± 3
Thana	bone	111 ± 2
Thana	brain	40 ± 2
	Indian shark (***Scoliodan sorrakowa***)	
Colaba	muscle	172 ± 8
Colaba	bone	318 ± 12
Thana	muscle	160 ± 6
Thana	bone	389 ± 14
Thana	brain	98 ± 4
	Chandawa (***Scatophagus argus***)	
Thana	muscle	327 ± 14
Thana	bone	392 ± 17
Thana	brain	88 ± 3
	Beckti, Khajuri (***Lates calcarifer***)	
Colaba	muscle	279 ± 9

continued

Table 7.7 (continued) Mercury in Various Varieties of Fish from the Colaba (Bombay) and Thana Markets

Market	Tissue	Hg Content (ng/g*)
Colaba	bone	382 ± 16
Thana	muscle	374 ± 8
Thana	bone	412 ± 12
Thana	brain	97 ± 4
	Cat-fish (***Macrones vitattus***)	
Thana	muscle	233 ± 5
Thana	bone	304 ± 11
Thana	brain	79 ± 1
	Black-fish (***Tilapia mozambique***)	
Colaba	muscle	243 ± 6
Colaba	bone	297 ± 8
Thana	muscle	417 ± 15
Thana	bone	444 ± 17
Thana	brain	102 ± 4

* fresh weight basis

c) INDIAN COASTAL FISH

Somayajulu and Rama (1972) have analysed the edible protein of seven varieties of fish collected from the coastal waters of Bombay, India. Their results (table 7.7) indicate that all of the analysed samples, except salmon, contained a mercury concentration of about 100ng/g (fresh weight basis). The tail portion of the salmon was found to contain 470 ± 40ng/g of mercury. This is very close to the permissible limit (500ng/g) in the USA and Canada. Tejam and Halder (1975) analysed thirty species of fish collected from the coastal waters of the industrialised regions of Bombay and Thana. Neutron activation techniques were used in this investigation in order to estimate mercury in the muscle, bone, and brain of fish. Their results are summarised in tables 7.7 to 7.10.

Table 7.8 Mercury in Various Varieties of Fish from Gujarat Fisheries, Colaba, Bombay (Tejam & Halder, 1974)

Tissue	Hg Content (ng/g*)
	Rohu (***Labeo rohita***)
muscle	122 ± 5
bone	164 ± 6
brain	91 ± 3
	Palla (***Hilsa ilisha***)
muscle	161 ± 6
bone	183 ± 7
brain	93 ± 3
	Catala (***Catala***)
muscle	193 ± 7
bone	212 ± 8
brain	104 ± 4
	fresh-water Shark (***Wallago attu***)
muscle	203 ± 8
bone	261 ± 9
brain	111 ± 4
	* fresh weight

Mercury concentrations in the brain, muscle, and bone of the various varieties of fish decreased in the order:

Brain - Muscle - Bone

Of the various types of fish collected, ***Harpodan nehereus*** had the lowest concentration of mercury while ***Hilsa ilisha*** had the highest concentration of mercury (about 500ng/g and about 375ng/g in bone and in muscle, respectively). The bones of ***Tilapia mozambique***, ***Lates calcarifer***, ***Wallago attu***, and ***Mugil dussumieri*** from the Membra creek contained 621ng/g, 610ng/g, 609ng/g, and 570ng/g, respectively, and their

Table 7.9 Mercury in Various Varieties of Fish from Murdeswar Lake (Thana) and Siddheswar Lake (Thana) (Tejam & Halder, 1975)

Lake	Tissue	Hg Content (ng/g*)
	Rohu (***Labeo rohita***)	
Murdeswar	muscle	165 ± 5
Murdeswar	bone	191 ± 8
Murdeswar	brain	73 ± 3
Siddheswar	muscle	202 ± 9
Siddheswar	bone	268 ± 11
Siddheswar	brain	89 ± 3
	Mrigal (***Cirrhina mrigala***)	
Murdeswar	muscle	132 ± 4
Murdeswar	bone	183 ± 6
Murdeswar	brain	64 ± 3
Siddheswar	muscle	168 ± 6
Siddheswar	bone	240 ± 10
Siddheswar	brain	80 ± 2
	common Carp (***Cyprinus carpio***)	
Murdeswar	muscle	171 ± 6
Murdeswar	bone	201 ± 5
Murdeswar	brain	80 ± 3
Siddheswar	muscle	260 ± 10
Siddheswar	bone	307 ± 12
Siddheswar	brain	94 ± 3
	Jitadi (***Telapia mozambique***)	
Murdeswar	muscle	230 ± 4
Murdeswar	bone	261 ± 6
Murdeswar	brain	98 ± 4

continued

Table 7.9 (continued) Mercury in Various Varieties of Fish from Murdeswar Lake (Thana) and Siddheswar Lake (Thana) (Tejam & Halder, 1975)

Lake	Tissue	Hg Content (ng/g*)
Siddheswar	muscle	305 ± 12
Siddheswar	bone	379 ± 15
Siddheswar	brain	115 ± 5
	Repsa (*Cirrhina reba*)	
Murdeswar	muscle	160 ± 5
Murdeswar	bone	189 ± 9
Murdeswar	brain	77 ± 3
Siddheswar	muscle	215 ± 7
Siddheswar	bone	292 ± 8
Siddheswar	brain	98 ± 4

* fresh weight

muscles contained 566ng/g, 462ng/g, 416ng/g, and 370ng/g, respectively. The levels of mercury in the bones of ***Tilapia mozambique***, ***Elops saurus***, ***Pellona motius***, ***Cyprinus carpio***, and ***Mugil dussumieri*** from the Thana creek were 603ng/g, 531ng/g, 520ng/g, 511ng/g, and 505ng/g, respectively. Since the permissible limit of mercury in fish from fresh and coastal waters was 500ng/g, the authors suggested that the consumption of ***Tilapia mozambique***, ***Lates calcarifer***, ***Wallago attu***, and ***Mugil dussumieri*** from the Mumbra creek, and ***Tilapia mozambique*** and ***Elops saurus*** from the Thana creek, should be restricted.

The total mercury contents in marine fish from the Indian Ocean have been presented by Kureisky et al (1979).

MERCURY CONCENTRATION IN THE ORGANS OF FISH

Mercury is distributed differently in the various organs of fish, and the maximum concentration is found in the tissues. Mercury levels in the muscle tissues of preserved museum fish which had been removed from the St Clair and Eire lakes between 1920 and 1965 were determined by Evanset al (1972).

Table 7.10 Mercury in Various Varieties of Fish from Mumbra Creek and Thana Creek (Tejam & Halder, 1975)

Creek	Tissue	Hg Content (ng/g*)
	Khajuri (***Lates calcarifer***)	
Mumbra	muscle	462 ± 11
Mumbra	bone	610 ± 15
Mumbra	brain	101 ± 4
Thana	muscle	390 ± 11
Thana	bone	414 ± 14
Thana	brain	98 ± 2
	Elops saurus	
Mumbra	muscle	392 ± 15
Mumbra	bone	481 ± 16
Mumbra	brain	104 ± 3
Thana	muscle	415 ± 15
Thana	bone	531 ± 10
Thana	brain	126 ± 5
	Mudadeshi (***Sillago sihama***)	
Mumbra	muscle	294 ± 11
Mumbra	bone	408 ± 7
Mumbra	brain	91 ± 2
Thana	muscle	266 ± 12
Thana	bone	401 ± 17
Thana	brain	84 ± 3
	Boi, Boity (***Mugil dussumieri***)	
Mumbra	muscle	376 ± 15
Mumbra	bone	570 ± 16
Mumbra	brain	108 ± 4

continued

Table 7.10 (continued) Mercury in Various Varieties of Fish from Mumbra Creek and Thana Creek (Tejam & Halder, 1975)

Creek	Tissue	Hg Content (ng/g*)
Thana	muscle	366 ± 16
Thana	bone	505 ± 23
Thana	brain	100 ± 4
	Pellona motius	
Mumbra	muscle	401 ± 16
Mumbra	bone	507 ± 18
Mumbra	brain	118 ± 5
Thana	muscle	411 ± 18
Thana	bone	520 ± 16
Thana	brain	112 ± 5
	Dhoma (***Sciaena dicanthus***)	
Mumbra	muscle	200 ± 8
Mumbra	bone	315 ± 11
Mumbra	brain	82 ± 1
Thana	muscle	217 ± 8
Thana	bone	347 ± 14
Thana	brain	94 ± 3
	fresh-water Shark (***Wallago attu***)	
Mumbra	muscle	416 ± 12
Mumbra	bone	609 ± 24
Mumbra	brain	106 ± 3
Thana	muscle	342 ± 7
Thana	bone	388 ± 11
Thana	brain	107 ± 5

continued

Table 7.10 (continued) Mercury in Various Varieties of Fish from Mumbra Creek and Thana Creek (Tejam & Halder, 1975)

Creek	Tissue	Hg Content (ng/g*)
	common Carp (***Cyprinus carpio***)	
Mumbra	muscle	309 ± 6
Mumbra	bone	477 ± 17
Mumbra	brain	109 ± 3
Thana	muscle	389 ± 15
Thana	bone	511 ± 21
Thana	brain	127 ± 5
	Jitadi (***Tilapia mozambique***)	
Mumbra	muscle	566 ± 22
Mumbra	bone	621 ± 31
Mumbra	brain	165 ± 5
Thana	muscle	512 ± 20
Thana	bone	603 ± 29
Thana	brain	148 ± 5

* fresh weight

These levels, compared to those in recently caught fish from the same areas, indicated that the mercury content had increased since the early 1940's, with an accelerated increase since the late 1960's.

Using flameless atomic spectrophotometry, Annett et al (1972) analysed 79 fish collected from 37 sediments of the Raisin River. The average total mercury content in the muscle tissue of several species of fish was 0.06 to 1.7ppm (wet weight basis).

The total mercury concentrations in the sediments were 0.19 to 0.53ppm (dry weight basis).

Krocza et al (1975) found that 96% of 106 fish (salmonids and Coregonids) caught in the Salzkammergut lakes contained less than 0.2ppm of Hg in their muscle tissues. No concentrations greater than 0.4ppm were found, even when 34 predatory fish and 47 cyprinids were included in the count. However, of the fish (cyprinids and three pike) caught in the Danube river and its tributories, only 18% had muscle mercury concentrations of less than 0.2ppm, 41% contained 0.2 to 0.5ppm, and 41% contained more than 0.5ppm. The four highest concentrations ranged from 1.6 to 1.9ppm.

A total of 1227 ***H. Stenolepis*** were analysed in order to determine the mercury content in the edible muscle tissue (Hall et al, 1976). These fish were obtained from 5 geographical areas: the Bering Sea, the Gulf of Alaska, south-east Alaska, British Columbia and Washington (Oregon). Mercury was found to be uniformly distributed from nape to tail in the edible muscle tissue. Within each area, the mercury concentration increased as the size of the fish increased. The mercury concentration also increased, in fish of a given size, from the north to the south of the species range.

The maximum mercury levels in the lateral muscles of northern pike (***Esox lucius***), burbot (***Lota lota***), and walleye (***Stizostedion vitreum***) from north-western Ontario were 27.8, 24.8, and 19.6ppm, respectively, in specimens taken 50 to 60 miles downstream from a chlorine plant (Fimreite & Reynolds, 1973). The mercury levels usually decreased with increasing distance from the plant, but high values continued to be detected as far as 200 miles downstream. They decreased in proportion to the distance from the plant but were elevated, even 200 miles downstream. The levels in specimens from presumably uncontaminated lakes were generally less than 1ppm but frequently greater than 0.2ppm, a value which was considered to represent the maximum background concentration.

The mercury content of the edible muscle tissue of swordfish which were captured in the straits of Messina ranged from 0.44 to 3.6ppm, with a mean of 1.65ppm.

The chronic effects of mercuric chloride intoxication upon the digestive system of ***Channa punctatus*** have been investigated in India (Shastry & Gupta, 1978). Marked histological changes were noted in the liver, including glycogen depletion and cirrhosis.

MERCURY LEVELS IN FISH FROM VARIOUS ENVIRONMENTS

Mercury contamination of fish in various waters, e.g. sea, lake, river, etc, has been reported, and mercury levels have been determined by various workers using various analytical methods. Tarao et al (1976) analysed Red sea-bream (***Chrysophys major***), mejina (***Girella punctat***), and marbled rock-fish (***Sebasticas marmoratus***) which had been reared for up to 45 days in sea-water containing sediments from Minamata Bay. The muds contained mercury (100 and 600ppm; dry weight basis) mainly in the form of insoluble compounds. Kikuchi et al (1976) analysed the organs of sea-bream after exposure to $HgCl_2$ (0.001, 0.01 and 0.1ppm) in sea-water for up to 13 weeks. The distribution of mercury decreased in the order:

spleen - brain - liver - muscle

The concentrations of mercury were 10 and 0.3ppm respectively in the spleen and muscle of fish.

Barbar et al (1972) reported the mercury contents of bottom-dwelling deep-sea fish such as ***Aldrovandia macroclion***, ***Antimora rostrata***, ***Bathysaurusagassizi***, ***Chalinuura brevibarbis***, and ***C. carpina*** which live at a depth of 2500m. These species had similar feeding habits but their mercury concentrations differed by an order of magnitude. One large (52.2cm long) specimen of ***A. rostrata*** contained approximately 0.74 to 0.76ppm of mercury.

Fish from Lake St Clair and western Lake Eire exhibited generally lower mercury contents. Among western Lake Eire fish, the gizzards had the lowest mercury content and walleyes the highest content.

Kim et al (1976) analysed fresh-water fish (carp, crucian carp, and Snake-head) from the Han River (Korea) using non-destructive neutron activation techniques, and found them to contain 0.031 to 0.63ppm of Hg, 0.83 to 0.11ppm of Se, and small amounts of 10 other trace elements. The mercury contents varied as a function of local sources of water pollution, and an interdependence of the levels of pairs of some elements was observed. The mercury concentration (together with that of 11 other trace elements) in anthracite samples from several mines was also determined in order to monitor possible pollution caused by their use. The values ranged from 0.84 to 1.3ppm. Mercury concentrations in the liver were higher than those in muscle (Calapaj et al, 1974). Of the muscle tissue from three different body locations, highly vascularised tissue contained the most mercury, and fat-rich tissue the least. Essentially constant mercury concentrations were found in the eggs, regardless of the weight of the fish or the mercury content of other tissues.

The total mercury content of some fish and shellfish along the Mexican coast were determined by Reimer and Reimer (1975).

The values ranged from 0.01 to 0.61 $\mu g/g$ in fish from Veracruz and Coatzacoalcos. In all of the fish investigated, the mercury content of the liver was higher, and sometimes more than double that in the muscle. The mercury levels in shrimp were low, except for the samples from Mazotlan and Guaymas. The contents in molluscs were very low, except in the case of ***Codakia orbicularis*** where one third of the levels were above 0.10ppm.

The average MeHg levels were 0.11 and 0.20 μg/g respectively in the liver and muscle of yellow-fin Tuna (*Thunnus algacares*) from the mid-Pacific, west Pacific, and east Indian oceans (Ueda & Takeda, 1977). Concentrations of MeHg in the dark muscle (0.21 μg/g) were slightly higher than those in the dorsal muscle (0.19 μg/g). The ratios of MeHg to total mercury content were 76.0, 87.5 and 84.6 respectively in the dorsal muscle, dark muscle and liver. However, Takeda et al (1976) found that the total mercury levels were 0.21, 0.24, and 0.25 μg/g respectively in the abdominal, dark, and dorsal muscle of yellow-fin Tuna from the mid-Pacific and east Indian oceans. The total mercury level in the muscles, liver (0.13 μg/g), stomach (0.12 μg/g), and intestine contents (0.05 μg/g) increased with length. No significant correlation was found between the total polar and non-polar lipid contents and the total mercury content in the dorsal and dark muscles.

The concentration of mercury in the muscle of Tuna fish (*Thunnus obseus*, *T. thunnus*) and swordfish (*Xiphias gladius*), as analysed by Arima and Umaemoto (1976), ranged from 0.74 to 2.34ppm. The highest distribution was found in the myofibrillar protein fraction, followed by the sarcoplasmic protein, and the soluble and non-protein fractions. From 57 to 68% of the muscle mercury was in the myofibrillar protein. A linear relationship was noted between the mercury content in the muscle and that in the myofibrillar protein or in the sarcoplasmic protein.

The accumulation of inorganic and organic mercury by the thornback ray, *Raja clavata L.* was studied by Pentreath (1976a,b,c) using $^{203}HgCl_2$ and $CH_3^{203}HgCl$. Observations were made of the rates of intake and ^{203}Hg loss from both labelled food and sea-water, and the internal distributions of the isotope were compared with that of the total mercury. Both forms were readily absorbed from sea-water. However, retention of the two forms from food was dissimilar in that, in contrast to inorganic mercury, methylmercury was readily absorbed and slowly eliminated. The results are compared with data on the accumulation of mercury by plaice.

MERCURY IN FISHERY PRODUCTS

Reviews on mercury in fish, shellfish, and fishery food products have been published by the Ministry of Agriculture (1971), Holden (1973), and Schreiber (1983). The relationship between the mercury contents in indigenous species, and the geology, industrial activities, and biology (migration), etc, have been discussed by Bernhard and Renzoni (1977) and Bernhard and Buffoni (1982).

Fish from the Mediterranean are known to contain a higher amount of mercury (Stoeppler et al, 1979a,b) as has been demonstrated for mackerel and blue-fin tunny (Bernhard & Renzoni, 1977; Renzoni et al, 1978).

Fish from the English Channel, southern part of the North Sea, German Bight, and Irish Sea contain higher levels of mercury as compared with those in fish from the northern part of the North Sea, or the Baltic.

The concomitant occurrence of selenium with mercury results in a reduction in the toxic effects of the latter. For marine fish, Se to Hg molar ratios greater than unity are considered to be characteristic of a natural marine environment (Freeman et al, 1978; Luten et al, 1980; Suzuki et al, 1980).

Results for striped mullet caught in Mediterranean waters show that the sum of mercury and selenium (as molar quantities) increases in proportion to the age of the fish. This is a strong indication of the presence, in this fish, of a so-called 'Se-Hg compartment' which is gradually filled up (Leonzio et al, 1982). The selenium contents in fish have been presented by several workers (Bebbington et al, 1977; Ueda & Takeda, 1977; Freeman et al, 1978; Itano et al, 1979; Luten et al, 1980; Cappon & Smith, 1981).

REFERENCES

- Abdulla, M.T., Royle, L.G., and Morris, A.W. (1972) - Nature, 235, 158-60
- Ackroyd, D.R. and Marsh, J.G. (1981) - referred to by Bland et al (1982)
- Aichberger, K. (1977) - Zeitschrift für Lebensmittel Unterschung und Forschung, 163(1) 35-8
- Alberta Interdepartmental Committee on Pesticides (1970) - Report prepared for the Minister, Alberta Department of Agriculture, Alberta, USA, 19th January, 1970
- Anderlini, V.C., Connors, P.G., Risebrough, R.W., and Martin, J.H. (1972) - Proceedings of a Symposium on Conservation Problems in Antarctica, 49-63
- Annett, C.S., Fadow, M.P., and D'Itri, F.M. (1972) - Michigan Academician, 4(3), 325-37
- Arima, S. and Umaemoto, S. (1976) - Swisan Gakkaishi, 42(8), 931-7
- Bache, C.A., Gutenmann, W.H., and Lisk, D.J. (1971) - Science, 172, 951

- Barber, R.T., Vijayakumar, A., and Cross, F.A. (1972) - Science, 178, 636-9
- Baun, C.E., Adrian, W.J., and Keiss, R.E. (1977) - Journal of Wildlife Management, 41(1), 131-4
- Bebbington, G.N., MacKay, N.J., Chvojka, R., Williams, R.J., Dunn, A., and Avty, E.H. (1977) - Australian Journal of Marine and Freshwater Research, 38, 277-86
- Beeftink, W.G. and Nieuwenhuize, J. (1982) - Science of the Total Environment, 25, 199-223
- Beides, A. and Higgons, D. (1957) - Journal of Science of Food and Agriculture, 8, 597
- Berg, W., Johnels, A., Sjostrand, B., and Westermark, T. (1966) - Oikos, 17, 71-83
- Berglund, F. and Berlin, M. (1969) - Chemical Fallout, Charles G.Thomas, Springfield, Illinois
- Berglund, F.M., Berlin, G., Birke, V., Von Euler, L., Friberg, L., Homstedt, E., Jonsson, C., Ramel, S., and Skerfving, A. (1971) - Report from an Expert Group, Nordisk Hygienisk Tidskrift, Supplement
- Berlin, M. and Ullberg, S. (1963) - Archives of Environmental Health, 6(5), 602
- Berlin, M., Fazacherley, J., and Nordberg, G. (1969) - Archives of Environmental Health, 18(5), 719
- Bernhard, M. and Buffoni, A. (1982) - Proceedings of an International Conference on Environmental Pollution, September, 1981, Thessaloniki, 458-84
- Bernhard, M. and Renzoni, A. (1977) - Thalassia Jugoslavica, 13, 265-300
- Bland, S., Ackroyd, D.R., Marsh, J.G., and Millward, G.E. (1982) - Science of the Total Environment, 22, 235-41
- Bol'shakov, A.P., Ptushku, L.I., and Sovel'eva, N.Y. (1970) - Biol. Rol. Mikroelem. Ikh, Primen, Sel-Khoz' Med., 1, 98-9
- Bongers, L.H. and Khattak, M.N., (1972) - USEPA Publication No. 16080 HVA, 245
- Bourne, W.R.P. (1973) - Marine Pollution Bulletin, 4, 77-9
- Bryan, G.W. (1976) - in Marine Pollution, (Edited by R.Johnston), Academic Press, London, 185
- Bundesgesundheitsblatt (1979) - Deutsches Lebensmittel Rundschau, 75, 319
- Burrows, E.E. and Combs, B.D. (1958) - Progressive Fish Culturist, 20, 143

- Buzadzhi, A. (1965) - Sb. Nauch. Tr., Donkoi Sel' skhokhoz. Inst., 2(1), 48

- Calapaj, R., Ciraolo, L., Calabro, G. (1974) - Guad Merceol, 13(1), 5.19 (Italy)

- Cappon, C.J. and Smith, J.C. (1981) - Archives of Environmental Contamination and Toxicology, 10, 305-19

- Chan Chen-Chang, Ke-Chin-Nau, Lo. Jiunn. Ghuang (1977) - HO Tzuk'O, Hsuch, 14(1), 9-12 (Chinese)

- Chen, R.W., Lacy, V.L., and Whanger, P.D. (1975) - Research Communications in Chemical Pathology and Pharmacology, 12, 297-308

- Chichoszowa, E.S., Hordynska, S., and Legatowa, B. (1969) - Przemyslowy Instytut Sadow, 13, 121-31 (Polish)

- Connors, P.G., Anderlini, V.C., Risebrough, R.W., Gilbertson, M., and Hays, H. (1975) - Canadian Field-Naturalist, 89, 157-62

- Corral, J. and Masso, C. (1975) - Boletin del Instituto Espanol de Oceanografiya, 184, 15pp (Spanish)

- Cumber, H., Curtis, E.H., and Blaylock, B.G. (1978) - Environmental Pollution, 17, 311-21

- Cunningham, P.A. and Tripp, M.R. (1975) - Marine Biology, 31, 321-34

- Dale, F.H. (1970) - Pesticide Regulation Division, US Agricultural Research Series, March 3, 1970

- Dale, I.M., Baxter, M.S., Bogan, J.A., and Bourne, W.R.P. (1973) - Marine Pollution Bulletin, 4, 77-9

- DeLisle, C.E. (1975) - International Conference on Heavy Metals in the Environment, Toronto, Canada, 821

- Eisler, R., Zaroogian, G.E., and Hennekey, R.J. (1972) - Journal of the Fisheries Research Board of Canada, 29, 1367-9

- Ekman, E. (1969) - Nordisk Hygienisk Tidskrift, 50(2), 116

- El-Begearmi, M.M., Sunde, M.L., and Ganther, H.E. (1977) - Poultry Science, 56, 313-22

- Ellis, R. and Fang, S. (1967) - Toxicology and Applied Pharmacology, 11(1), 104

- El-Sokkary, I.H. (1982) - Science of the Total Environment, 23, 55-60

- Evans, R.J., Bails, J.D., and D'Itri, F.M. (1972) - Environmental Science and Technology, 6(10), 901-5

- Fimreite, N. (1974) - Journal of Wildlife Management, 38, 120-31

- Fimreite, N. and Reynolds, L.M. (1973) - Journal of Wildlife Management, 37(1), 62-8

- Fimreite, N., Brun, E., Froslie, A., Frederichsen, P., and Gunderson, N. (1974) - Astarte, 7, 71-5

- Flaschentrager, E. (1951) - Physiologische Chemie, I, 208

- Forbes, R., Cooper, A., and Mitchell, H. (1954) - Journal of Biological Chemistry, 209, 857

- Forney, R. (1950) - Federation Proceedings of the American Society for Experimental Biology, 8, 292

- Freeman, H.C., Shum, G., and Uthe, J.F. (1978) - Journal of Environmental Science and Health, A13, 235-40

- Friberg, L. (1956) - Acta Pharmacologia et Toxicologica, 12, 411

- Friedman, M.A., Eaton, L.R., and Carter, W.H. (1978) - Bulletin of Environmental Contamination and Toxicology, 19, 436-43

- Furutani, S. and Osajima, Y. (1965) - Kyushu Daigaku Nogakubu Gakgai Zasshi, 22(1), 45

- Furutani, S. and Osajima, Y. (1967) - Nippon Shokohiu. Kogyo Ghakkaishi, 14(1), 15

- Gambrell, R.P., Collard, V.R., Reddy, C.R., and Patrick, W.H. (1977) - Technical Report D-77, Department of the Army, Washington, DC, 123

- Ganther, H.E., Goudie, C., Sunde, M.L., Kopecky, M.J., and Wagner, P. (1972) - Science, 175, 1122-4

- Gilbertson, H. (1974a) - Bulletin of Environmental Contamination and Toxicology, 12, 726-32

- Gilbertson, M. (1974b) - Canadian Field-Naturalist, 88, 273-80

- Goldschmidt, V.M. (1954) - Geochemistry, Oxford University Press, 730

- Goldwater, L., Ladd, A., and Jakobs, M. (1964) - Archives of Environmental Health, 9(4), 735

- Griffith, G., Bath, E., and Walker, J. (1954) - Annals of Internal Medicine, 41, 501

- Griffith, W.H. (1971) - Workshop on Mercury in the Western Environment, Portland, Oregon, USA, 25-26th February, 1971

- Gustavsson, I. and Golimowski, J. (1981) - Science of the Total Environment, 22, 85-7

- Gutenmann, W. and Lisk, D. (1960) - Journal of Agricultural and Food Chemistry, 8, 306-8

- Hall, A.S., et al (1976) - Fisheries Bulletin, 74(4), 783-9

- Hamilton, G. and Ruthven, A. (1967) - Journal of Science of Food and Agriculture, 18(12), 558

- Hannerz, L. (1968) - Institute of Fresh Water Research, Drottningholm, Sweden, Report No. 48, 120

- Harper, W.G. (1960) - Proceedings, Washington, US Public Health Ser. 181

- Harris, R.C., White, D.B., and Macfarlane, R.B. (1970) - Science, 170, 736

- Hasselrot, T.B. (1968) - Institute of Fresh Water Research, Drottningholm, Sweden, Report No. 48, 102

- Hauser, E. and Mohadjerani, H. (1976) - Mitteilungen aus dem Gebiete der Lebensmitteluntersuchung und Hygiene, 67(3), 389-401

- Hawkes, H.E. and Webb, J.S. (1962) - Geochemistry in Mineral Exploration, Harper and Row, New York, 415

- Heisinger, J.F., Hansen, C.D., and Kim, J.H. (1979) - Archives of Environmental Contamination and Toxicology, 8, 279-83

- Helper, L.G. and Olofsson, G. (1975) - Chemical Reviews, 585-602

- Holden, A.V. (1973) - Journal of Food Technology, 8, 1-25

- Huckabee, J.W., Carton, F.O., and Kennington, G. (1972) - Journal of Wildlife Management, 36(4), 1306-9

- Hughes, H.L. (1957) - Annals of the New York Academy of Sciences, 65, 454-60

- Itano, K., Sasaki, K., and Akehashi, H. (1979) - Journal of the Food Hygiene Society of Japan, 20, 299-306

- Jacobs, M. (1964) - Archives of Environmental Health, 9(4), 454

- Jacobs, M. and Goldwater, L. (1961) - Food Technology, 15(5), 357

- Jernelov, A. (1970) - Limnology and Oceanography, 15, 958-60

- Jernelov, A. (1971) - Vatter, 26(3), 262 (Translation)

- Johnels, A.G., Westermark, T., Berg, W., Persson, P.I., and Sjostrand, B. (1967) - Oikos, 18, 323

- Johnels, A.G. and Westermark, T. (1969) - in Chemical Fallout, Charles G.Thomas, Springfields, Illinois, 221

- Karppanen, E., Henrikson, K., and Helminen, M. (1970) - Nordisk Medicin, 84(35), 1097-9

- Karppanen, E., Henrikson, K., and Nurmi, E. (1968) - Annales Agriculturae Fenniae Supplements, 7(1), 24-9

- Kikuchi, T., Honda, H., Morinchi, H., and Amano, K. (1976) - Shokuhiu Eissigaku Zasshi, 17(6), 438-43

- Kim, N., Lee, C., and Bak, H. (1976) - Journal of the Korean Nuclear Society, 8(1), 41-7

- Klaverkamp, J.F., Turner, M.A., Harrison, S.E., and Hesslein, R.H. (1983) - Science of the Total Environment, 28, 119-28

- Koeman, J.H., Van den Ven, W.S.M., de Goeji, J.J.M., Tjioe, P.S., and Van Haaften, J.L. (1975) - Science of the Total Environment, 3, 279-87

- Kramer, H.J. and Neidhart, B. (1975) - Bulletin of Environmental Contamination and Toxicology, 14, 699

- Krocza, W., Haide, W., Weidenholzer, W. and Stoeckl, W. (1975) - Tierdererztl Monaterschrift, 62(11), 405-10

- Kudo, A. and Mortimer, D.C. (1979) - Environmental Pollution, 19, 239-45

- Kureisky, T.W., George, M.D., and Sen Gupta, R. (1979) - Marine Pollution Bulletin, 10, 357-60

- Kyle, J.H. and Ghani, N. (1983) - Science of the Total Environment, 26, 157-62

- LAC Annual Report (1980) - Verslag van de Werkzaamheden van de Landbouwadvies-commissie Milieukritische Stoffen in 1980, Ministry of Agriculture and Fisheries, The Hague, 153pp

- Larsson, J.E. (1970) - Swedish Environmental Protection Board Research Secretariat

- Legatowa, B., Hordynska, S., and Kaszyca, A. (1970) - Proz. Asstw. Saki, 21-7

- Leonzio, C., Focardi, S., and Bacci, E. (1982) - Science of the Total Environment, 24, 249-54

- Levitan, S., Rosner, L., and Yannai, S. (1974) - Israeli Journal of Zoology, 23(3-4), 135-42

- Lofroth, G. (1969) - Working Group on Environmental Toxicology, Ecological Research Committee of the Swedish Natural Science Research Council, Stockholm, Sweden, 38pp

- Loughton, A. and Richard, F. (1974) - Mushroom Science (Pub. 1976) 9 pt. 1, 347-7

- Lundgren, K., Swensson, A., and Ulfvarson, U. (1967) - Scandinavian Journal of Clinical Laboratory, 20(2), 164

- Luoma, S.N. (1977) - Marine Biology, 41, 269-73

- Luoma, S.N. (1983) - Science of the Total Environment, 28, 1-22

- Luten, J.B., Ruiter, A., Ritskes, T.M., Rauchbaar, A.B., and Riekwel-Booy, G. (1980) - Journal of Food Science, 45, 416-9

- Mackay, N.J., Kazacos, M.N., Williams, R.J., and Leedow, M.I. (1975) - Marine Pollution Bulletin, 6, 57-61

- Martin, J. and Pickard, J. (1957) - Annual Report of the Agricultural and Horticultural Research Station, Long Ashton, Bristol, 76

- Martoja, R. and Martoja, M. (1978) - L'Antagonisme Selenium - Mercure, Actualités de Biochimie Marine, Colloque GABIM, La Rochelle, 149-84

- Ministry of Agriculture, Fisheries, and Food (1971) - Survey of Mercury in Food, Working Party on the Monitoring of Foodstuffs for Mercury and Other Heavy Metals, 1, Report, HMSO, London

- Nadkarni, R.A. and Ehman, W.D. (1971) - Radiochemistry and Radioanalytical Letters, 6(7), 89-96

- Nixon, G., Paxton, G., and Smith, H. (1967) - Journal of Dental Research, 44, 654

- Noren, K. and Westöö, G. (1970) - Fisheries Board of Canada, Translation Series No. 135, from Var Foods, 2(13), 1967

- Norseth, T. (1967) - Biochemical Pharmacology, 16(9), 1645

- Norseth, T. (1968) - Biochemical Pharmacology, 17, 2788

- Novick, S. (1969) - Environment, 11, 2

- Ohlendorf, H.M., Risebrough, R.W., and Vermer, K. (1978) - US Federation of W.S. Wild Research Report, 9, 1-40

- Ohno, T., Kishida, H., and Shimizu, N. (1967) - Gifu Yakka Daigaku Kijo, 17, 88

- Pallotti, G., Bencivenga, B., and Vegliante, A. (1976) - Industria Alimenticia, 15(12), 57-60 (Italy)

- Peden, J.D., Crothers, J.H., Waterfall, C.E., and Beasley, J. (1973) - Marine Pollution Bulletin, 4, 7-9
- Pentreath, R.J. (1976a) - Journal of Experimental Marine Biology and Ecology, 24, 121
- Pentreath, R.J. (1976b) - Journal of Experimental Marine Biology and Ecology, 24, 103-19
- Pentreath, R.J. (1976c) - Journal of Experimental Marine Biology and Ecology, 25, 51
- Perkins, J.W. (1971) - Geology Explained in South and East Devon, David and Charles, Newton Abbot, 196
- Phillips, G.R. and Buhler, D.R. (1978) - Transactions of the American Fisheries Society, 107, 853-61
- Phillips, G.R. and Gregory, R.W. (1979) - Journal of the Fisheries Research Board of Canada, 36, 1516-9
- Pickard, J. and Martin, J. (1960) - Journal of Science of Food and Agriculture, 14(10), 706
- Platonow, N. (1968) - Occupational Health Review, 20(1-2), 1
- Rankama, K. and Sahama, T.G. (1950) - Geochemistry, Chicago University Press, 912
- Rauter, W. (1975) - Zeitschrift für Lebensmittel Unterschung und Forschung, 159(3), 149-51 (German)
- Reimer, A.A. and Reimer, R.D. (1975) - Bulletin of Environmental Contamination and Toxicology, 14(1), 105-11
- Reith, J. and Gerritsma, K. (1955) - Voding, 16, 832
- Renzoni, A., Bernhard, M., Sara, R., and Stoeppler, M. (1978) - Journées d'Etude de Pollution Anatyla, CIESM, 225-60
- Ruzicka, J. and Lamm, C. (1969) - Talanta, 16(2), 157
- Sahagian, B. (1967) - Journal of Nutrition, 93, 291
- Sahagian, B., Harding Barlow, I., and Perry, H. (1966) - Journal of Nutrition, 90, 259
- Salei, P.I. (1970) - Veterinarva (Moscow), 47(8), 102-3
- Schreiber, W. (1983) - Science of the Total Environment, 31, 283-300

- Shacklette, H.T. (1965) - US Geological Survey Bulletin, 1198-c, 18
- Shacklette, H.T. (1970) - US Geological Survey Professional Paper, 713, 35
- Shastry, K.V. and Gupta, P.K. (1978) - Environmental Research, 16, 270
- Shitara, K. and Yasumasa, A. (1976) - Hokkaidoritzu Eigel Kenkyushoho Hokkaidorita, 26, 73-8
- Shultz, C.D., Greer, D., Pearson, J.E., Rivers, J.B., and Hylin, J.W. (1976) - Bulletin of Environmental Contamination and Toxicology, 14, 230-4
- Sims, R.R. and Presley, B.J. (1976) - Bulletin of Environmental Contamination and Toxicology, 16, 520
- Sivic, M., der Marderosian, A., Ullicci, P., and Fenical, W. (1974) - Food Drugs, Sea. Proc. 4th., 145-51
- Smart, N. and Hill, A. (1968) - Journal of Science of Food and Agriculture, 19, 315
- Sollman, M. and Schreiber, N. (1936) - AMA Archives of Internal Medicine, 57, 46
- Somayajulu, B.L.K. and Rama (1972) - Current Science, 44(6), 207-8
- Staarink, T. and Hakkenbrak, P. (1982) - Het Contaminantenboekje. Overzicht van Stoffen die eet-en Drinkwaren Verontreinigen, Den Haag, Staatsuitgeverij, 92
- Stewart, C. and Stolman, A. (1960) - Toxicology I, Academic Press, New York, 206
- Stock, A. (1940) - Biochimische Zeitschrift, 304, 73
- Stock, A. and Cucuel, F. (1934) - Naturwissenschaften, 22, 390-3
- Stoeppler, M., Bernhard, M., Backhaus, F., and Schulte, E. (1979a) - Science of the Total Environment, 13, 209-23
- Stoeppler, M. and Nürnberg, H.W. (1979b) - Ecotoxicology in Environmental Safety, 3, 335-51
- Stoneburner, D.L. and Harrison, C.S. (1981) - Science of the Total Environment, 19, 51-8
- Suzuki, N. (1959) - Science Reports of Tohoku University, First Series, 43, 161
- Suzuki, T., Satoh, H., Yamamoto, R., and Kashiwazaki, H. (1980) - Bulletin of Environmental Toxicology, 24, 805-11
- Takeda, M., Inamasu, Y., Koshilkawa, T., Ueda, T., Nakano, M., Tomida, T., and Hamada, M. (1976) - Suisan Daigakko Kenkyu Hokoku, 25(F), 47-65
- Takeda, Y. (1968) - Toxicology and Applied Pharmacology, 13(2), 165

- Tankawa, Y. (1969) - Hokkaidoritzu Eigel Kenkyushoho, 19, 117

- Tarao, R., Tabata, T., and Yasuhara, M. (1976) - Nippon Suisan Gakkaishi, 42(12), 1411-22 (Japan)

- Tejam, B.M. and Halder, B.C. (1975) - Indian Journal of Environmental Health, 17, 9

- Tejning, S. (1967) - Rapport 670206, Yrkesmedicinska Kliniken via. Lasarettet i Lund

- Thomas, A. (1976) - Fette Scifen Austrich, 78(4), 141-4

- Thornton, I., Watling, H., and Darracott, A. (1975) - Science of the Total Environment, 4, 325

- Ueda, T. and Takeda, M. (1977) - Nippon, Suisan Gakkaishi, 43(9), 1115-21

- Ukeles, R. (1962) - Applied Microbiology, 10, 532

- Ulfvarson, U. (1965) - Svensk Khemi Tidskrift, 77(5), 235

- Vermeer, K. (1971) - paper presented at the 35th North American Wildlife and Natural Resources Conference, Portland, Oregon, USA, 7-10th March, 1971

- Wallace, R.A., Fulkerson, W.D., and Lyon, W.S. (1971) - Mercury in the Environment, Oak Ridge National Laboratory.

- Warren, H.V. and Delavault, R.E. (1969) - Oikos, 20, 537

- Wester, P. (1964) - Aktiebolaget Atom Energie AE-169, Stockholm, Sweden, 20

- Westermark, T., Odsjo, T., and Johnels, A.G. (1975) - Ambio, 4, 87-92

- Westöö, G. (1969) - in Chemical Fallout, Charles G.Thomas), Springfield, Illinois

- Woidich, H. and Pfannhauser, W. (1975) - Deutsches Lebensmittel Rundschau, 71(5), 177-8 (German)

- Yasumasa, K. (1968) - Eisei Kagaku, 14(2), 76

- Zingmark, R.G., Knowles, S.C., and Miller, T.G. (1975) - Plant Aiken., South Carolina, Report SRO-777-1 (pts 1-3), 76pp

CHAPTER VIII

MERCURY POLLUTION: POISONING AND THERAPY

INTRODUCTION

Mercury is an unique pollutant because of its apparent indestructibility. Like lead, it does not play an essential role in the vital functions of plants and animals and, like lead, is an ubiquitous element. The atmosphere at ground level near to mercury ore deposits may contain as much as $20000 mg/m^3$ of Hg in air. On a weight basis, this amounts to almost 16lb of mercury per billion pounds of air. The maximum value in air is found at mid-day, with much smaller amounts being found in the morning or evening. The saturation level of mercury in air in equilibrium with metallic mercury increases exponentially with increasing temperature (Vaughn, 1967). The various sources of mercury to the environment, e.g. mining and extraction operations for mercury winning from nature, chlor-alkali, paper, paint, fungicide, and electrical industries, dental and other amalgams, and instruments such as DC meters, pressure gauges, etc., have been discussed in an earlier chapter. Other sources are the burning of fossil fuels (coal, petroleum), sulphide ores, etc. These industries contribute increasingly to pollution with the progress of industrialisation. The efforts of society to check this increase will be discussed later.

Mercury appears in industrial discharges in five principal forms: (1) divalent mercury, Hg^{2+}, (2) metallic mercury, Hg°, (3) Phenylmercury, $C_6H_5Hg^+$, (4) alkoxyalkyl mercury, $CH_3O\text{-}CH_2\text{-}CH_2\text{-}Hg^+$, and (5) methylmercury, CH_3Hg^+. These compounds can be ranked in order of decreasing toxicity thus: methylmercury compounds, mercury vapour, inorganic salts of mercury and a number of organic forms such as phenylmercury salts. Methylmercury, the most toxic form, is formed mainly by methylation of mercury by the methanogenic bacteria which are widely distributed in the sediments of rivers and canals and in the sludge of sewage beds. The nature of mercury pollution and poisoning of the human system have been discussed in detail by Mitra and Ghosh Dastidar (1976).

Mercury fungicides are ingested directly by mammals (e.g. small rodents) by consuming treated seeds, or possibly by eating the sprayed stems and leaves. Predators can contract mercury poisoning indirectly by eating contaminated animals (birds or rodents). In the human system, mercury fungicides are absorbed via the lungs, skin, or stomach. The heavy treatment of rice plants with PMA in Japan has increased the mercury level of the Japanese people approximately three-fold with respect to the average for other populations (Fukunaga & Tsukano, 1969). The mercury level in the blood of those working in seed-dressing factories is higher than average (Tejning, 1967). The half-life for the elimination of methylmercury in man is 75 days, and that of inorganic mercury is about 42 days (Rahola et al, 1971). Hundreds of deaths have resulted from the inadvertent ingestion of treated cereal seed by humans. In Sweden, three people were poisoned by eating porridge made from treated cereal seed (Engelson & Herner, 1952). Several hundred people were poisoned, resulting in many deaths, in northern Iraq in 1956, due to eating home-made bread made from mercury-treated cereal seed (Jalili & Abbasi, 1961). In Pakistan, more than one hundred people were poisoned, with many fatalities, due to the ingestion of bread made from wheat seed treated with ethyl and phenylmercury compounds (Haq, 1963). Many peasant families in Guatemala were similarly poisoned by methylmercury (Eyl, 1971).

METALLIC MERCURY AND INORGANIC SALTS

Mercury has long been recognized as being a general cellular poison and effective protein precipitant. In the elemental form, mercury can contaminate air either by evaporation from metal droplets or from the solution of mercury salts (particularly those of strong oxyacids) which are highly ionised because the $Hg^{\circ} + Hg^{2+} \rightleftharpoons 2Hg^{+}$ equilibrium ensures the continuous formation of Hg°.

Chronic mercurialism may appear after a few weeks of exposure, or it may be delayed for longer periods. Psychological and emotional disturbances are characteristic. Neurological disturbances also appear. Stomatitis is a common manifestation. There has been a long controversy over the correlation of urinary mercury levels and mercurialism. It appears that the mercury level in urine is important with regard to the early exposure of individuals, but may be of no value for monitoring long-term exposure. Mercury in saliva can be of diagnostic importance in the case of mercurialism (El-Sadik et al, 1970).

Table 8.1 Acute Toxicity of Inorganic Mercury Salts. (IAEA, 1972, p70)

Compound	Species	Route	LD_{50}*	Reference
mercurous iodide	mouse	oral	110	Gothe & Sundell, 1964
mercurous iodide	rat	oral	310	Gothe & Sundell, 1964
mercuric acetate	rat	oral	104	Ikeda, 1968
mercuric iodide	mouse	oral	80	Gothe & Sundell, 1964
mercuric iodide	rat	oral	40	Gothe & Sundell, 1964
mercuric chloride	rat	S.C.	3	Clarkson, 1968
mercuric chloride	mouse	S.C.	6	Clarkson, 1968

*(mg/kg body-weight)

Mercury vapour is almost completely absorbed through the alveolar membrane (Kudsk, 1965; Vostal & Clarkson, 1973) and is oxidized in the blood and tissues before reacting with biologically important binding sites. The time interval between the passage of elemental mercury through the alveolar membrane and the time required for complete oxidation is long enough (Magos, 1967, 1968) to produce a more pronounced accumulation in the central nervous system than after the ingestion of inorganic mercury salts (Magos, 1968; Berlin et al, 1966). The oxidation time can be prolonged by compounds such as ethanol (Magos et al, 1973) or aminotriazole (Magos et al, 1974).

After the acute administration of inorganic salts to animals and man, the highest levels of inorganic mercury are found in the kidneys and the second highest levels in the liver. The brain accumulates mercury to a greater extent when exposed to vapour than when exposed to organic mercury salts (Report of an International Committee, 1969).

In man, the excretion of inorganic mercury occurs via the kidneys, the liver (as bile), the intestinal muscosa, the sweat glands, and the salivary glands. However, the urinary and faecal routes of excretion are the most important. Elimination from the brain, thyroid, and testes is slow. Thus, accumulation of mercury in these organs is possible (Report of an International Committee, 1969).

Table 8.2 Acute Toxicity of Arylmercury Compounds

Compound	Species	Route	LD_{50}*	Reference
phenylmercuric acetate	mouse	oral	13	Piechocka & Kranze, 1967
phenylmercuric acetate	rat	oral	60	Piechocka & Kranze, 1967
phenylmercuric acetate	mouse	oral	70	Goldberg et al, 1950
phenylmercuric acetate	rat	oral	46	Ikeda, 1968
phenylmercuric acetate	chicken	oral	~60	Miller et al, 1960
phenylmercuric acetate	rat	i.p.	~10	Swensson, 1952
phenylmercuric nitrate	mouse	i.v.	27	Wein, 1939

*(mg/kg body-weight)

The administration of N-acetyl-Dz-penicillamine by mouth was associated with both a clinical improvement and increased urinary excretion (Brugsch, 1965). Another mercurial ailment called acrodynia pink disease is characterised by fever accompanied by a pink coloured rash, swelling of the spleen and glands, hyperkeratosis and oedema of the fingers (Magos, 1975). This disease has been virtually eliminated by the withdrawal of calomel as an cathartic or antihelminthic agent.

The acute toxicity of inorganic mercury salts and the lethal concentrations of mercury compounds for some organisms are given in tables 8.1 and 8.2, respectively.

ORGANIC MERCURY COMPOUNDS

Many of the environmental pollutants are in organic form but, in living tissue (animals and fish; Westöö, 1969), most of the mercury pollutant is present as methylmercury. The mechanisms of conversion of inorganic mercury to methylmercury will be discussed later. Phenylmercury compounds in seed-dressing agents can be converted to methylmercury compounds following initial conversion to inorganic mercury (Miller et al, 1960; Westöö, 1967).

Several groups of workers investigated the nature of the retention and elimination of mercury from animals injected with phenylmercuric compounds (Miller et al, 1960;

Swensson & Ulfvarson, 1968; Innes et al, 1969; Berlin, 1963; Fitzhugh & Nelson, 1950; Ichikawa, 1970; Kluge et al, 1938; Piechocka, 1968)
The acute toxicity of arylmercury compounds is summarised in table 8.2 (IAEA, 1972, p71).

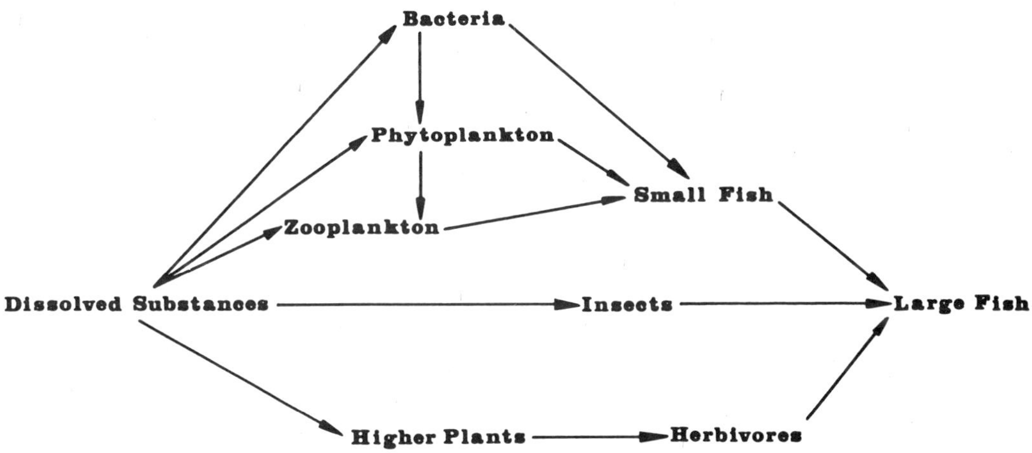

Figure 8.1.1 Diagrammatic Representation of the Flow of Mercury through an Aquatic Food Chain.

Chronic human exposure via the inhalation of phenylmercuric compounds at levels of up to 5.1mg of mercury per m^3 of air have not resulted in any evidence of clinical injury (von Massman, 1957; Ladd et al, 1964; Goldwater et al, 1964).

BIOTRANSFORMATION

The causative factor in the incidence of Minamata disease was traced to the pollution of the bay by the mercury-containing effluent from a large chemical plant. The isolation of methylthiomethylmercury from shellfish in the Minamata area led to the suggestion that mercury could be alkylated by "plankton and other marine life". Later, it was discovered that the spent catalyst of a reactor, which was discharged into the sea, contained about 1% methylmercury. It was thought that the biological transformation of mercury to methylmercury was quite insignificant.
In Sweden, when certain bird populations decreased drastically, investigation revealed the presence of large amounts of mercury in the dead birds. Furthermore, increased

Figure 8.1.2 Aquatic Food Chain. (A: anaerobic bacteria, B: birds, D: dissolved substances, I: insects, L: large fish, M: man, P: plankton, S: small fish, T: detritus. At each trophic level, less mercury is excreted than is ingested. Thus, there is proportionately more mercury in algae than in the water in which they live, yet more in fish that feed upon the algae, etc. Bacteria, and decay, promote the conversion of any mercury present into methylmercury.)

amounts of mercury were observed in fish as well as in people consuming large quantities of fish caught in polluted waters. The death of the birds was related to the extensive use of methylmercury dicyanodiamide as a fungicide in Swedish agriculture. The use of methylmercury compounds as fungicides has been banned in Sweden and, since then, the mercury content of birds and eggs has decreased substantially. The mercury in fish was found to be entirely in the form of methylmercury, and this was surprising in view of the fact that the fish had been caught in lakes and rivers into which no discharge of methylmercury had occurred. Jensen and Jernelov (1969) provided the first indication that the alkylation of mercury may be bacterial in origin, and that an aerobic ecosystem can affect the reaction. They have shown that mercury is methylated by aquarius sediments and by micro-organisms. These findings provided a conclusive explanation of the fact that methylmercury is found in fish even if all of the mercury discharges consist entirely of inorganic mercury. Diagrammatic representations of the flow of mercury through an aquatic food chain are shown in figures 8.1.1 and 8.1.2.

In a biological system, three agents can transfer methyl groups: S-adenosyl-methionine, N^5-methyl tetrahydrofolate derivatives, and methylcorrinoids (Wood et al, 1971). The first two transfer methyl groups as carbonium ions (CH_3^+) and are therefore incapable of methyl transfer to Hg^{2+}. However, the methylcorrinoids could generate methyl groups as carbanion ions (CH_3^-), carbonium ions (CH_3^+), or radicals ($CH_3^\bullet$). Therefore, methylcorrinoids methylate inorganic Hg salts (with CH_3^-) or Hg metal (with $CH_3^\bullet$) (Vallee & Ulmer, 1972)

The transformation of mercury to methylmercury can be considered to be a detoxification reaction from the point of view of bacteria since it frees them of mercury (Jensen & Jernelov, 1969; McBirde & Edwards, 1977) and this has been substantiated by the finding that mercury-tolerant strains of ***N. Crassa*** have a higher methylating capacity than do normal strains (Landner, 1971). Wood et al (1968) have suggested that mercury enters the bacterial cells in the form of Hg^{2+} and is then converted to Hg°; the conversion being favoured by the low redox potential at which these anaerobic organisms thrive. Dimethylmercury is then formed by methyl radical addition (Wood et al, 1968). After its formation, dimethylmercury may diffuse out of the cell and, at alkaline pH values, would be released into the water or, at acidic pH values, might possibly be converted to monomethylmercury and methane (Vallee & Ulmer, 1972). Chemical methylation of inorganic mercury with methylcobalamin, a vitamin B_{12} analogue, has been demonstrated by Imura et al (1971).

The biosynthesis of methylmercurials in aquatic environments, and its toxicological effects, have prompted studies of the metabolic and chemical cycles of mercury (Jensen & Jernelov, 1967, 1968; Wood et al, 1968,). They have confirmed that dimethylmercury is the primary product of the biological methylation of Hg; monomethylmercury being its decomposition product (figure 8.2).

As well as micro-organisms, animals are also capable of converting mercury to methylmercury within their bodies (Westöö, 1967, 1968) although, in the case of fish, it is not known whether fish themselves can convert Hg^{2+} to methylmercury. Indeed, several aerobic, facultative, and anaerobic bacteria, as well as mammalian liver, contain a cobalamin-dependent N^5-methyltetrahydrofolate homocysteine transmethylase (Guest et al, 1960; Burke et al, 1970) which is capable of the synthesis of methylmercury.

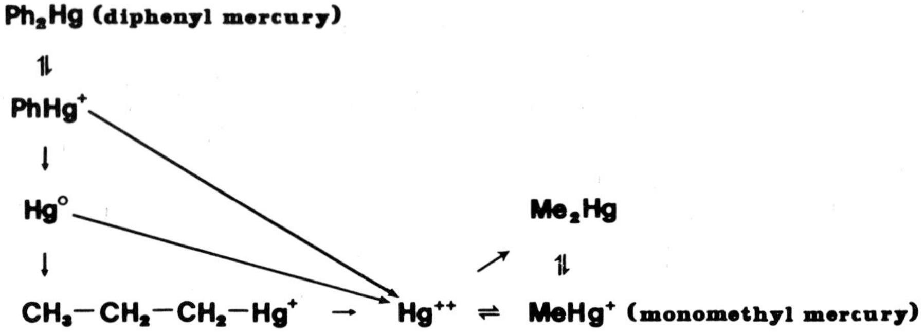

Figure 8.2 Steps by which Mercury and its Compounds are Converted to Methylmercury

A non-enzymatic pathway for the methylation of mercury can also work. Alkylation of inorganic mercury in the sludge pit of a caustic soda factory was reported by Yamaguchi et al (1971). Non-enzymatic synthesis by methylcobalamin has also been suggested by Hill et al (1970) and Vallee & Ulmer (1972). The reaction proceeds by the electrophilic attack of Hg^{2+} on methylcobalamin.

Microbial methylation progresses either by (1) enzymatic methylation or (2) non-enzymatic transfer of methyl groups by methylcobalamin to Hg^{2+}. The bacterially mediated synthesis of methylmercury is cyclic and methylmercury synthesis may be a mechanism for detoxification in bacteria exposed to inorganic mercury in sediments (Hamdy et al, 1977). Mercury in sediments may be remobilised by the bacteria via methylmercury synthesis and/or degradation and volatilisation, or by reducing inorganic mercury to a more volatile form. In San Francisco Bay, a higher concentration of methylmercury has been found in deeper sediments. Degradation of methylmercury proceeds at a faster rate in aerobic sediments (Olson & Cooper, 1976).

Mercury in fish and other aquatic organisms is mainly in the form of methylmercury (Westöö, 1966). Micro-organisms and methanogenic bacteria in sediments can methylate inorganic mercury (Jenson & Jernelov, 1967; Wood et al, 1968). Chemical and biological processes are involved in the methylation, as shown in figure 8.3.

Methylation takes place under anaerobic as well as aerobic conditions. In sulphur-bearing reducing media, HgS and HgS_2^{2-} form and prevent methylation.

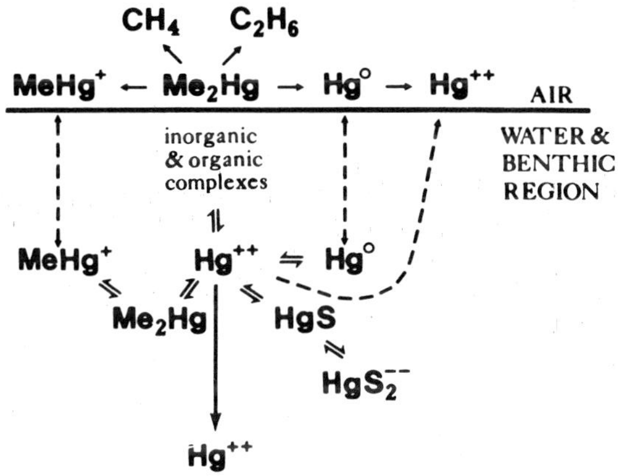

Table 8.3 Mercury Transformation in Air and Aquatic Environments

Akagi et al (1972) and Kitamura and Sumino (1972) reported that inorganic mercury is photomethylated by UV radiation, or in the dark, by using such compounds as methyl donors. Rogers (1976, 1977) showed that mercuric ions were further methylated in the NaOH solution extract of an agricultural soil, and suggested that the unidentified methylating factors were low molecular weight organic compounds which existed in the fulvic acid fraction of the extract. Both humic acid and fulvic acid, obtained from leaf mould and river sediments, methylate inorganic mercury. Methylation occurs in the dark and in bacteria and is also influenced by the reaction temperature, inorganic Hg concentration, humic acid and fulvic acid concentrations, and the pH of the solution (Nagase et al, 1982).

Bacteria also possess the ability to decompose organic mercury compounds (Suzuki et al, 1968; Tonomura & Kanazaki, 1969a,b). Cell-free extracts of a mercury-resistant ***Pseudomonas*** catalyse the decomposition of phenylmercuric acetate into benzene, thereby releasing volatile mercury (Tonomura & Kanazaki, 1969a,b).

Organomercurials are reduced by UV radiation (Inoko, 1981). Methylmercury (halide) can also be decomposed by UV radiation (without the formation of dimethylmercury):

$$CH_3HgI \xrightarrow{h\nu} CH_3 + HgI$$

Such a reaction might take place in the upper layer of an aquatic environment where light can penetrate to a reasonable extent.
Photodecomposition of dimethylmercury produces ethane and metallic mercury:

$$CH_3HgCH_3 + h\nu \rightarrow C_2H_6 + Hg$$

Organo-mercurials (such as dimethylmercury) in aqueous solution are finally transformed to metallic mercury and/or inorganic mercury compounds. Photodecomposition of methylmercury can occur via (Inoko, 1981):

$$CH_3HgCl + h\nu \rightarrow [CH_3 + HgCl] \rightarrow CH_3 + HgCl \begin{cases} 2CH_3 \rightarrow C_2H_6 \\ 2HgCl \rightarrow Hg_2Cl_2 \end{cases}$$

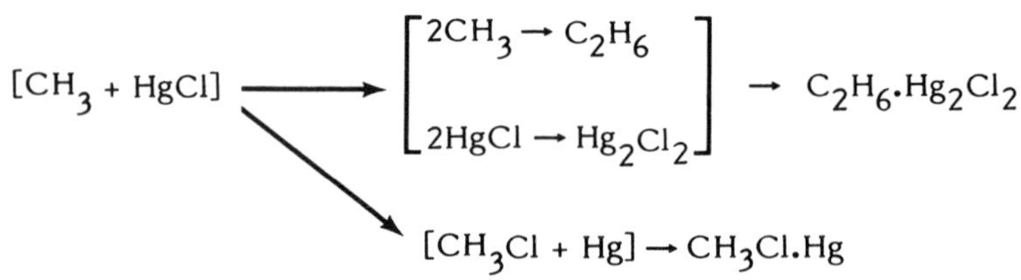

where [ ] is the three-dimensional cage of the H_2O molecule.

METHYLMERCURY

Methylmercury is many times more toxic to man than is the metallic form, and this is the form of mercury which is of concern with regard to pollution control. The earliest cases of poisoning were due to occupational exposure following the introduction of methylmercury compounds as antifungal seed dressing agents. Reports of poisoning from non-occupational sources appeared with increasing frequency from 1950 onwards. Methylmercury is more stable than other mercurials used as fungicides. Once it is bound to the seed it is very difficult to remove. Magos and Jawad (1973) have observed that even washing dressed wheat overnight in running water removed only 40% of the methylmercury.
The extent of exposure of the general population to methylmercury appears to involve

chiefly fish and possibly other foods, and not water or air directly. Methylmercury can be absorbed through the skin, respiratory tract, and alimentary canal (Berglund et al, 1971). Data from rats and man indicate that more than 90% of the methylmercury in food is absorbed (Berglund & Berlin, 1969). Methylmercury salts, if applied to the skin e.g. as an ointment, may give rise to clinical toxicity in man. There is considerable risk of absorption of methylmercury by the respiratory tract as a result of the occupational handling of seed dressings.

The toxic effects of methylmercury and its homologues are quite different to those resulting from the ingestion of other mercury compounds. The primary signs and symptoms of methylmercury poisoning result from damage to the nervous system. Methylmercury intoxication is characterised by ataxia (loss of coordination in gait), dysarthia (slurred speech), paresthesia (loss of sensation at the extremities of the hands and feet and around the mouth), tunnel vision (constriction of the visual field), and loss of hearing. Severe poisoning can cause blindness, coma, and death. The first symptoms, usually numbness in the extremities and in the perional area, do not usually appear until a few days after exposure. The mean latent periods often range from 16 to 38 days (Bakir et al, 1973). Although the conditions of the patients may improve somewhat, the damage to the nervous system is irreversible.

Approximately 10% of the total body burden of methylmercury in man is found in the head, most of it presumably being in the brain. Also, particularly high levels of methylmercury are found in the liver and kidneys (Berglund et al, 1971). Methylmercury accumulates in red blood cells and, in man, these cells contain over 90% of the methylmercury in the bloodstream (Report of an International Committee, 1969). Thus, the concentration of methylmercury in blood is the best indication of the body burden in people exposed to methylmercury. The concentration of methylmercury in milk, urine, cerebrospinal fluid and amniotic fluid is less than that in blood. The measurement of the concentration of methylmercury in consecutive segments of hair samples is the best means of recapitulating the history of exposure. A direct relationship between the levels of mercury in blood and hair was found for Swedes who were considered to have reached an equilibrium state between dietary intake and body burden of mercury. In fatal cases of Minamata disease, the ratio of mercury in brain, liver, and kidneys was 10:40:50 (Study Group on Mercury Hazards, 1971). The symptoms of methylmercury poisoning, and the resultant brain damage, are described accurately by Hunter and Russell (1954). The symptoms of methoxyethylmercury poisoning have been reported by Zeyer (1952).

The main route of excretion in man is via the faeces, where the rate is about ten times higher than that for urine. Elimination from the body can be described by a monophasal exponential function. The mean half-life of methylmercury in man is 70 days. There is a marked variability in the half-lives for clearance, and this indicates that individuals face different degrees of hazard from methylmercury. Those having the longest clearance half-lives will retain a higher mercury concentration for longer periods and probably face the greatest hazard. The factors responsible for the long biological half-life of methylmercury are its slow decomposition to organic mercury (Task Group on Metal Accumulation, 1973), its complete reabsorption when detected in bile (Norseth & Clarkson, 1971), and its low urinary excretion compared to inorganic mercury (International Committee, 1969). Because of the low elimination rate in man, the steady-state between uptake and elimination is reached approximately one year after exposure has started.

Methylmercury can affect genetic material. When added to a medium in which plant cells are growing, it causes disturbances in the mechanisms of cell division and chromosone breakage. It also causes disturbances in the mechanisms of cell division in fruit flies and in tissue cultures from mice and man (Berglund et al, 1971). In those who have been exposed to methylmercury via the consumption of fish, a correlation has been found between the mercury levels of blood cells and the frequency of chromosome breakage in circulating lymphocytes cultured in vitro (Berglund et al, 1971; Larsson, 1970). There are grounds for assuming that the effects acting upon genetic material in man are of the same character as those acting upon lower animals. Berglund et al (1971) state that it must be assumed that exposure to methylmercury involves certain genetic risks. However, it is not possible, on the basis of presently available data, to estimate the extent of such risks and their relationship to different exposures to methylmercury.

In experiments on animals, methylmercury has been shown to cause teratogenic effects. According to the report of an international committee (1969), data from the Minamata cases presented by Nurakami indicate that teratogenic effects occur at an earlier stage of development than would be the case for central nervous system damage from methylmercury intoxication. This report states that "because of the experimental evidence of strong effects of methylmercury compounds on cell division and chromosome segregation, it is conceivable that this early effect may have resulted from induced chromosomal alterations of humans".

Another point of significance is that methylmercury concentrates in the foetus, possibly leading to damage to newborn children, although the mother shows no symptoms of mercury poisoning. Methylmercury readily crosses from mother to foetus, the concentration in the blood of the newborn is equal to, if not higher than that in the mother, and the intake of contaminated maternal milk after exposure in utero counteracts the decline in concentration caused by excretion and growth. Infants exposed to mercury poisoning in the uterus face a greater hazard than those exposed to contaminated milk. The report of an international committee (1969) states that "studies in animals and men indicated that methylmercury easily penetrates to the foetus via the placenta. The concentration of mercury in the foetal blood is about 20% higher than in the mother and the same statement should apply to the brain of the foetus as well". In the Minamata incident, 22 children (5.8%) born between 1955 and 1958 had cerebral palsy. The expected frequency of cerebral palsy was 0.1 to 0.6% (Berglund et al, 1971). None of the mothers had clinical symptoms of methylmercury poisoning.

Berglund et al (1971) have concluded that, of 230 cases of known methylmercury poisoning, slightly more than 20 were prenatal. A total of 20 cases were due to occupational exposure, three cases were due to skin preparations containing methylmercury, and the rest were due to the consumption of food containing methylmercury, of which 150 were due to eating fish and shellfish, while the rest were due to eating seed treated with methylmercury or meat from animals which had been given such seed. One third of the cases of methylmercury poisoning in Minamata were lethal (Berglund et al, 1971). The reports of Hunter (1955) indicate that a 16 year old boy who was exposed for only a few months to methylmercury compounds was unable to work after 20 years because of persistent ataxia tremors and an inability to recognize objects by touch. The forty families affected by Minamata disease had 61 cats, of which 50% were affected by methylmercury poisoning (Kurland & Firo, 1960). Commonly observed symptoms in cats from the affected households included unsteadiness, frequent falls, circling movements, and convulsions. The sensitivity of human beings to methylmercury is presumably higher than that of cats, which is higher again than that of rats (Takeuchi, 1970).

The lowest reported level in the brains of people who had died of methylmercury poisoning was estimated to be about 5 μgHg/g at the onset of the disease

(Berglund et al, 1971). They estimated the total body burden in cases of fatal poisoning to be about 80mg of mercury. Assuming a daily elimination rate of 1% of the total body burden, they calculated that the body burden of 80mg is reached after prolonged exposure corresponding to 0.8mg of mercury per day. Berglund et al (1971) state that uncertain trace dose data indicate that about 1% of the total body burden is found in 1000ml of whole blood and, for a body burden of 80mg, this would correspond to about 0.8 μg/g of mercury in whole blood. They further state that, on the basis of available material, it would appear to be justified to assume that clinically manifested poisoning of adults sensitive to methylmercury may occur at a level in whole blood of 0.2 μg/g mercury. This is reached upon exposure to about 0.3mg of mercury per day or about 4 μg/kg body weight per day.

According to Berglund et al (1971): "A safety factor must be applied between the lowest mercury level and the exposure that can be assumed to cause clinically manifested intoxication with neurological symptoms in adults and an acceptable mercury level and exposure for the population". They were of the opinion that a factor of 10 gives a sufficient safety margin. Thus, acceptable levels of mercury would be as follows:

(1) whole blood - 0.02 μg/g, (2) red blood - 0.04 μg/g, and (3) hair - 6 μg/g.

Based upon the above levels, the acceptable daily intake of methylmercury would correspond to 0.03mg of mercury or approximately 0.4μg/kg body weight (Berglund et al, 1971).

SUMMARY OF TOXICITY

Tambon (1972) has summarised the salient points relative to the toxicity of methylmercury.

1 - Man absorbs from ingested food practically all of the methylmercury present.

2 - Methylmercury in man is relatively stable, i.e. it remains in man as methylmercury.

3 - The half-life of methylmercury in man is between 70 and 90 days.

4 - Because of the slow elimination rate in man, a steady state between uptake and elimination is reached approximately one year after exposure has started.

5 - Methylmercury has a propensity for the human nervous system, and about 10% of the total body burden is in the head, presumably most of it being in the

brain.

6 - Methylmercury is neurotoxic.

7 - Postnatal methylmercury poisoning is not easy to diagnose, especially in the case of mild or atypical symptoms.

8 - Apart from raised mercury levels in blood and hair, no clinical laboratory investigations have given any clear and common positive findings for methylmercury poisoning.

9 - Diagnosis of methylmercury poisoning is based upon neurological symptoms and it is conceivable that brain lesions may occur at lower exposures and levels than those which cause neurological symptoms or could be diagnosed using available methods.

10 - It is not possible, with presently available data, to estimate what the long-term effects of sub-clinical methylmercury poisoning are on man.

11 - In non-fatal cases of methylmercury poisoning, disability can persist for an extended period of time.

12 - Compensatory mechanisms of the nervous system can delay clinical recognition of methylmercury poisoning, even when brain damage has already occurred.

13 - There are presently no known drugs which are effective in the treatment of methylmercury poisoning.

14 - There are probably individual variations in sensitivity to methylmercury.

15 - Methylmercury is teratogenic.

16 - Prenatal methylmercury poisoning cannot be distinguished from other types of cerebral palsy, and diagnosis would have to be made epidemiologically with the supporting evidence of mercury levels in blood and hair.

17 - In man, concentrations of methylmercury in foetal blood are about 20% higher than those in the mother's blood.

18 - There is a greater risk of methylmercury poisoning to the foetus than to the mother, and affected children can be born to mothers showing no clinical symptoms of methylmercury poisoning.

19 - Methylmercury has been shown to be mutagenic to test organisms.

20 - A correlation in man, between the frequency of chromosome breakage in lymphocytes and the mercury level in blood cells, has been demonstrated.

21 - It must be assumed that exposure of man to methylmercury involves certain genetic risks. However, it is not possible, on the basis of presently available

data, to estimate the extent of such risks.

22 - An intake by man of 0.8mg per day of mercury (as methylmercury), corresponding to a level of 0.8mg/g of mercury (as methylmercury) in whole blood, may be fatal.

23 - Clinically manifested poisoning of adults sensitive to methylmercury may occur at a level in whole blood of 0.2mg/g mercury (as methylmercury). This can be reached upon exposure to about 0.3mg mercury (as methylmercury) per day.

24 - On the basis of point 23, and using a safety factor of 10, acceptable levels of mercury (as methylmercury) would be 0.02mg/g in whole blood and 6mg/g in hair.

25 - On the basis of point 24, an acceptable daily intake of mercury (as methylmercury) would be 0.03mg per day.

OTHER ALKYLMERCURY COMPOUNDS

The other forms of short-chain alkylmercury compounds are ethylmercury and diakylmercury. Ethylmercury is less stable than methylmercury and its relatively faster decomposition in vivo is shown by its somewhat different distribution and by the higher level of urinary excretion in animal experiments (Ulfvarson, 1962; Miller & Clarkson, 1973). Also, in cases of poisoning in man, the kidneys are frequently involved (Damluji, 1962). Ethylmercury is not synthesized in the ecosystem so the only risks of exposure occur during the production or consumption of dressed foodstuffs.

Diethylmercury is highly volatile and was the cause of death of two stenographers who worked in a room which was adjacent to a warehouse in which huge amounts of diethylmercury were stored (Hill, 1943). Dialkylmercury is less toxic than the corresponding monoalkylmercury salt because a significant part of the dose is converted to the monoalkylmercury salt (Ostlund, 1969). In occupational exposure, the maximum allowable concentration of short-chain alkylmercury compound is 100ng/ml blood (International Committee, 1969).

COMPOUNDS OTHER THAN ALKYL

Such compounds may enter the body by inhalation, skin absorption, or oral ingestion (Report of an International Committee, 1969). The report of another such committee (1965) states that the efficiency of absorption by the respiratory tract is unknown but should be high. The committee also reports that when mercury is present as an acrovol, it is likely to dissolve quickly in body fluids and be distributed in the blood. Methoxymethylmercury shows a distribution pattern in the body which is similar to that of inorganic mercury compounds, except that a higher concentration may be present in the liver, alimentary tract, and red blood cells, and a lower concentration present in the renal cortex (Report of an International Committee, 1969). Methoxymethylmercury is degraded in the body to inorganic mercury. The toxicity of organic mercury compounds is similar to that of mercuric salts (Report of an International Committee, 1969).

CLINICAL MANIFESTATIONS OF POISONING IN MAN

The usual symptoms of mercurial poisoning due to continuous or repeated exposure to mercury-polluted air in cities, industrial areas, and factories are, as discussed earlier:
Digestive disturbances such as nausea, loss of appetite, and colicky pain in the abdomen. The gums become tender, swollen, inflamed, and ulcerated (spongy gum). Salivation becomes profuse. The tongue and cheeks become swollen and ulcerated. The breath becomes foul and a blue line appears at the junction of gum and teeth, while the latter become loose in their sockets. Emaciation and diarrhoea begin. Skin eruptions may appear (eczymatous or erythematous in character). Mercurial tremor in the body develops. Speech becomes hesitant and stammering. The victim becomes progressively anaemic and slight mental derangements such as insomnia and delirium appear. Paralysis of the arms and legs may occur, and the patient dies of exhaustion.
The treatment of such cases will start with advice to leave the factory or mercury-polluted area. Administration of any mercury-containing drug is stopped. Gargling with alum and potassium chlorate or borax should be performed frequently. Sodium thiosulphate in doses of between 0.45 and 6g in 5cc of water is said to be very useful. A course of Dz-penicillamine (BAL) may be given.

THERAPY

This should involve the removal of the toxic metal from liver and other parts of the body by administering (orally or by injection) a chelating agent. Known chelates have a low molecular weight and a high polarity. They link to toxic (and non-toxic) metals and are filtered out of the blood by the kidneys. This is how they act so as to cause excretion of the metal in a chelating form. However, many chelating agents appear to facilitate the uptake of toxic metals by the kidneys, which are more vulnerable than the liver. The latter possesses some ability to repair damage, whereas the kidneys do not.

By increasing biliary excretion, the poisoning effect of metals such as Hg, Pb, and Cu may be reduced. However, heavy metal compounds (e.g. CH_3Hg^+) may subsequently be re-absorbed with the bile acids and participate in enterohepatic circulation. For this reason, the net excretion of such metal compounds via the faecal route may be inadequate.

Besides the kidneys, the liver is one of the principal storage sites for many metals. In a study by a group of doctors and biochemists (Jones et al, 1976), four approaches to the development of chelating (enterohepatic) agents suitable for biliary rather than renal excretion of toxic metal ions were developed.

The basic features exploited in the design of such enterohepatic chelating agents were those which (1) make the chelating agent large enough and sufficiently non-polar to prevent filtration, and/or (2) provide it with a structure which allows it to participate in the enterohepatic circulation of bile acids and their derivatives. The altered electric charge following metal ion chelation is the major factor preventing the continued enterohepatic recycling of the metal chelate complex. These agents promote faecal rather than urinary excretion of the toxic metals.

Of the four approaches suggested by these workers, the first is to use chelating agents of sufficiently high molecular weight (300 or more), and appropriately distributed polar and non-polar portions such that filtration through the glomerulus into the urine is hindered. In such cases, biliary excretion is then the preferred route (Smith, 1973). This behaviour becomes apparent when the molecular weight approaches 400. The agents which fall into this category are compounds such as 5,7-diiodo-8-hydroxyquinoline (diagrams I to III).

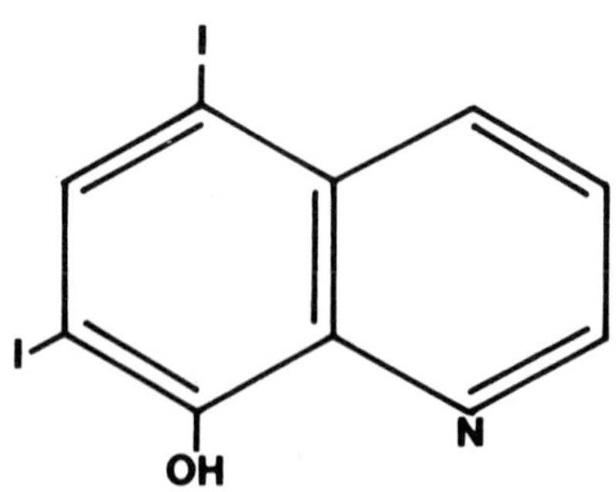

I

In the second approach, chelating agents with structural features similar to those of bile acids are used. In such compounds, the change in charge which occurs upon coordination to the metal ion should hinder re-absorption of the metal chelate complex and hence favour elimination of the metal via the faeces. The synthesis of such chelating agents (diagrams IV to VI) has been discussed by them.

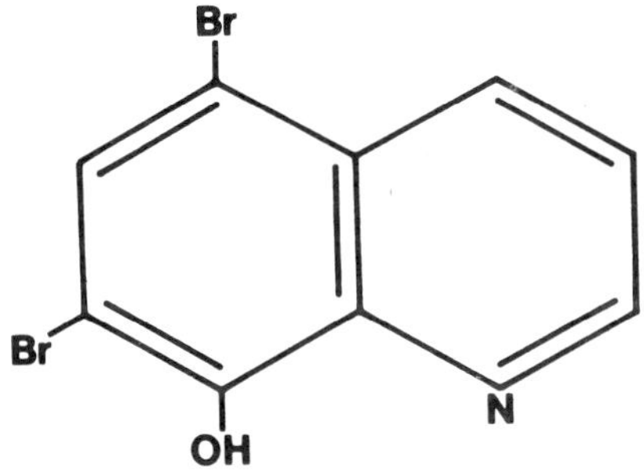

II

The third approach employs chelating agents which are structurally similar to fatty acids, but contain polar end-groups which are resistant to the normal metabolic processes which operate on fatty acids. Compounds of this type (diagrams VII to X) participate in enterohepatic circulation.

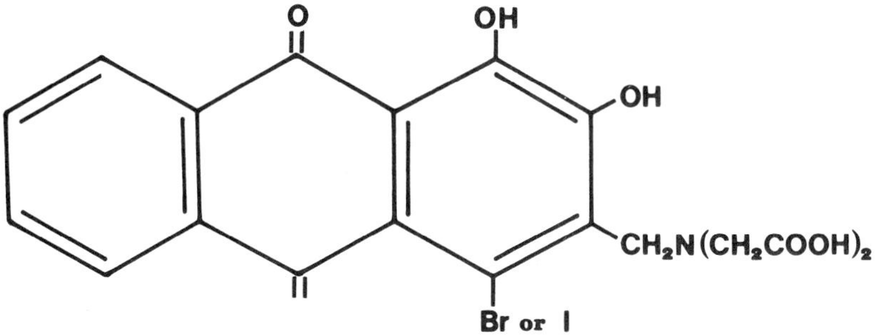

III

The fourth approach involves enhancing faecal excretion by the oral administration of an indigestible polymer containing chelating sites which is capable of removing toxic metal ions from enterohepatic circulation via the formation of stable complexes which pass on through the intestinal tract.

The last approach has been used in studies on mercury and has been found to be effective with derivatives of hair (Takakashi & Hirayama, 1971), a substituted polystyrene polymer containing sulphydryl groups (Clarkson et al, 1971; Bakir et al, 1973), and a mercaptal polymer specifically synthesized for that purpose (Jones & Harbison, 1974).

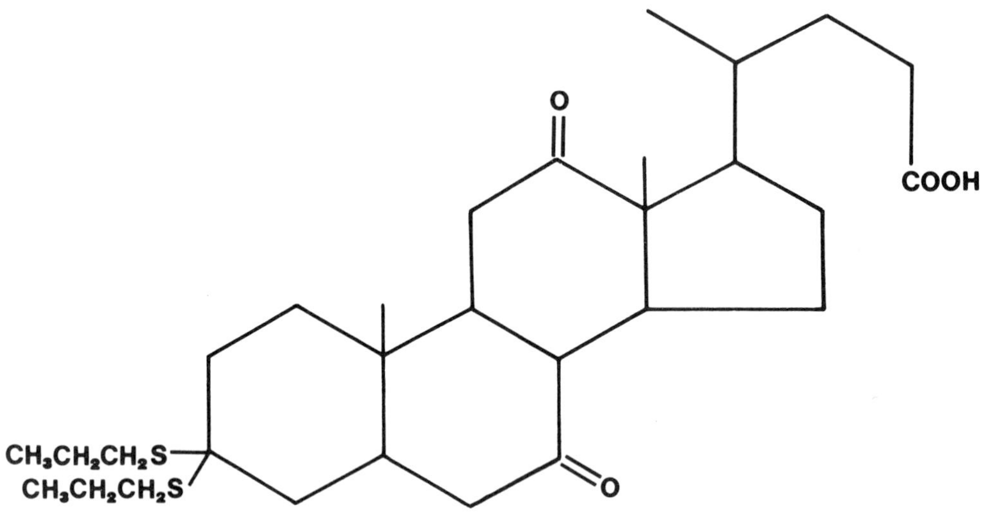

IV

The clearance of mercury from blood can be accelerated by the administration of Dz-penicillamine (N-acetyl-Dz-penicillamine, 2-3 dimercapto-propanol) also known as BAL. This compound is suitable for inorganic mercury poisoning but is ineffective in the case of methylmercury poisoning. A comparatively new addition to this group of compounds is a thiol resin. This has been found to be highly effective in enhancing the elimination of methylmercury compounds from experimental animals (Clarkson et al, 1973). The first administration of thiol resin to human beings was successfully carried out after the poisonings in Iraq and was found to possess two distinctive advantages over diffusable complexing agents which are distributed systematically. Firstly, it does not result in redistribution of mercury in the body and, secondly, it does not produce any toxic side effects (Bakir et al, 1973).

Recent research indicates that the biliary excretion of methylmercury in rats and

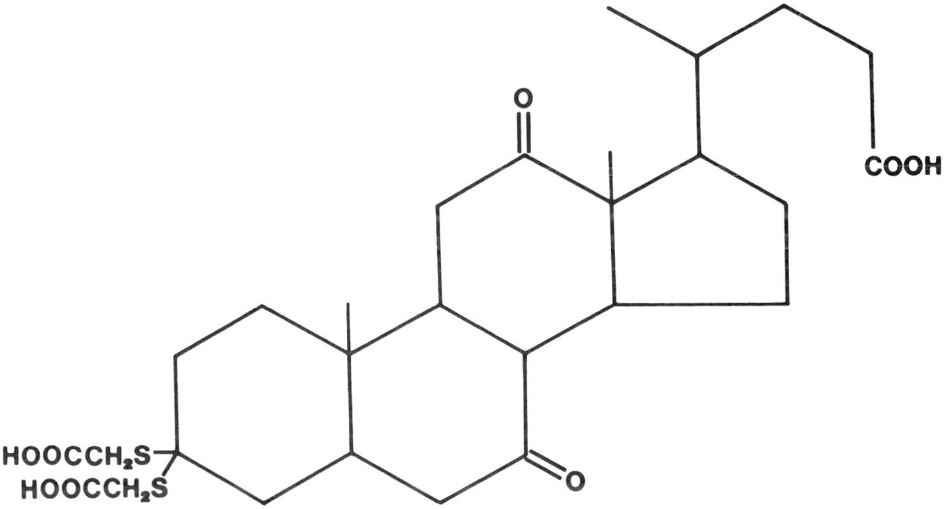

V (LD_{50} >500mg/kg)

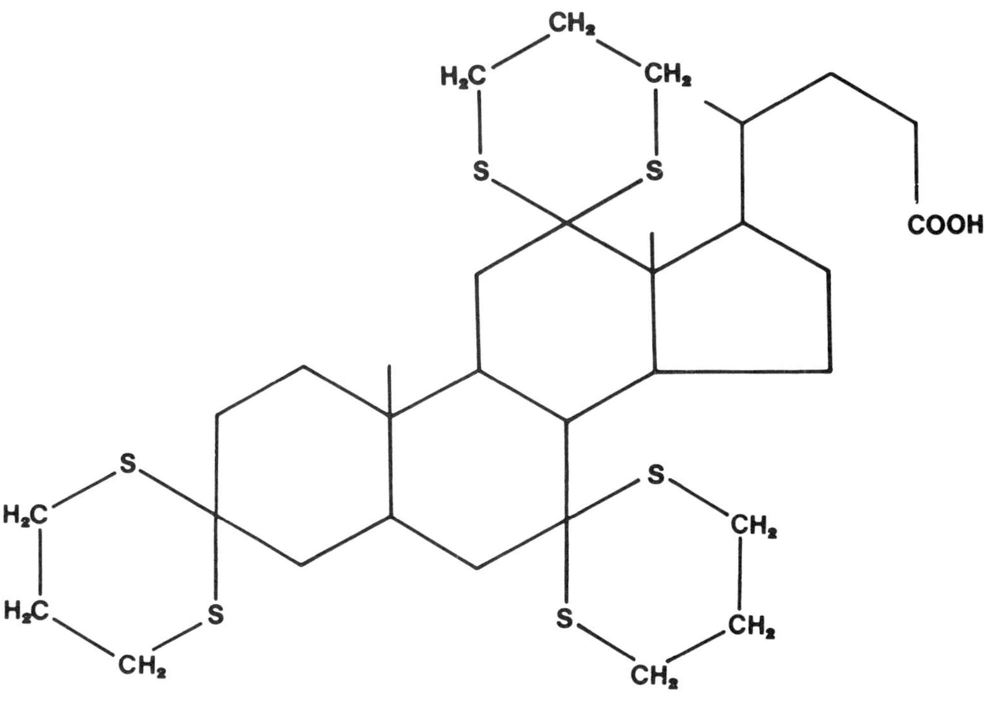

VI (LD_{50} >500mg/kg)

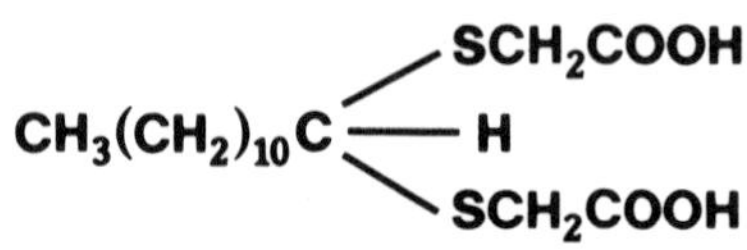

VII

mice can be increased by phenobarbitone which, combined with the administration of non-sorbable thiol resin to trap methylmercury, can significantly shorten the half-life of this substance (Magos & Clarkson, 1973).

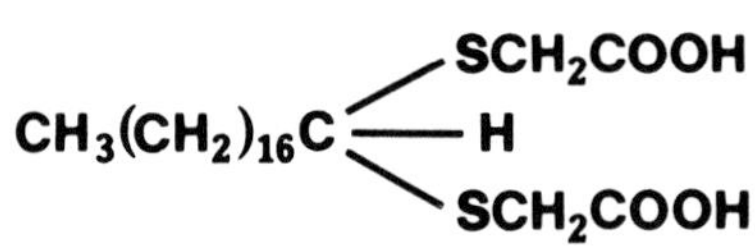

VIII

Studies of the mechanism of oxidation of mercury in human blood (Magos et al, 1974) indicated that peroxide-generating systems such as methylene blue or menadione accelerated mercury uptake in vitro, and that ethanol and cyanide were inhibitory.

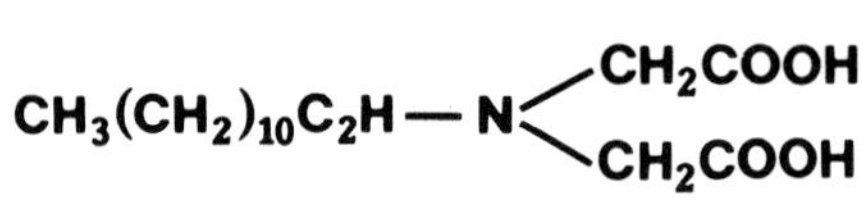

IX

Fang (1974) has reported that the activities of phenylmercuric acetate cleavage enzymes in the liver were increased in rats supplemented with 0.5 to 5ppm of sodium selenite in drinking water.

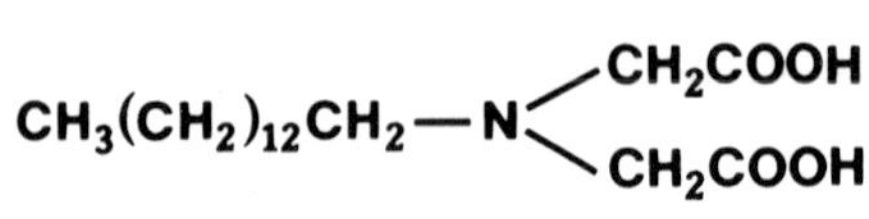

X

Stilling et al (1974) have shown that the addition of cysteine to diets reduced the toxicity to a marked extent. A combination of cysteine and selenium produced a considerably greater additive effect in rats.

REFERENCES

- Akagi, H., Fujita, M., and Sakagami, Y. (1972) - I. Eiseikagaku, 18, 309-14
- Bakir, F., Damluji, S.F., Amin-Zaki, L., Murtadha, M., Khalidi, A., Al-rawi, N.Y., Tikrite, S., Dhahir, H.I., Clarkson, T.W., Smith, J.C., and Doherty, P.A. (1973) - Science, 181, 230
- Berglund, F. and Berlin, M. (1969) - in Chemical Fallout, Charles C.Thomas, Springfield, Illinois
- Berglund, F., Berlin, M., Birke, G., Von Euler, V., Friberg, L., Eomstedt, B., Jonsoon, E., Ramel, C., Skerfring, S., Swensson, A., and Tejning, S. (1971) - Methylmercury in Fish, A Toxicological Epidemiological Evaluation of Risks. Report from an Expert Group, Chapter 1, Abstract. Nordisk Hygienisk Tidskrift, Supplement
- Berlin, M. (1963) - Acta Medica Scandinavica, Supplement No. 396
- Berlin, M. and Ullberg, S. (1963) - Archives of Environmental Health, 6, 589-601
- Berlin, M., Jerksell, L.G., and Von Ubisch, H. (1966) - Archives of Environmental Health, 12, 33
- Brugsch, H.G. (1965) - Journal of Occupational Medicine, 7, 394
- Burke, G.T., Mangum, J.H., and Brodin, J.D. (1970) - Biochemistry, 19, 4297
- CEK Report (1970) - Environmental Science and Technology, 4, 890
- Clarkson, T.W. (1968) - Journal of Occupational Medicine, 10, 351
- Clarkson, T.W., Small, H., and Norseth, T. (1971) - Feb. Proc., 30, 543
- Clarkson, T.W., Small, W., and Norseth, T. (1973) - Archives of Environmental Health, 26, 173
- Damluji, S.F. (1962) - Journal of the Faculty of Medicine, Baghdad, Iraq, 4, 83-102
- El-Sadik, Y. and Abdel-Aziz El-Dakha, M. (1970) - American Industrial Hygiene Association Journal, 31, 705

- Engelson, G. and Herner, T. (1952) - Acta Paediatrica, 41, 289
- Eyl, T.B. (1971) - Journal of the American Medical Association, 215, 287
- Fang, S.C. (1974) - Research Communications in Chemical Pathology and Pharmacology, 9, 579
- Fitzhugh, O. and Nelson, A. (1950) - Archives of Industrial Hygiene, 2, 5, 433ff
- Friberg, L., and Vostal, J. (1972) (Editors) - Mercury in the Environment: An Epidemiological and Toxicological Appraisal, CRC Press, Cleveland, USA
- Fukunaga, K. and Tsukano, Y. (1969) - Residue Review, 26, 1
- Goldberg, A.A., Shapiro, M., and Wilder, E. (1950) - Journal of Pharmacy and Pharmacology, 2, 20
- Goldwater, L.J., Ladd, A.C., and Jacobs, M.B. (1964) - Archives of Environmental Health, 9, 735
- Gothe, C.J. and Sundell, L. (1964) - Archives of Toxicology, 20, 226
- Guest, J.R., Helleiner, C.W., Cross, M.J., and Woods, D.D. (1960) - Biochemical Journal, 76, 396
- Hamdy, M.K., Noyes, O.R., and Wheeler, S.R. (1977) - in Biological Implications of Metals in the Environment, ERDA Symposium Series No. 42, 20-35
- Haq, I.U. (1963) - British Medical Journal, 1, 1579
- Hill, H.A.O., Pratt, J.M., Risdale, S., Williams, F.R., and Williams, R.J.P. (1970) - Chemical Communications, 124, 341
- Hunter, D. (1955) - Mercury in Diseases of Occupations, Boston, Little, Brown
- Hunter, D. and Russell, D.S. (1954) - Journal of Neurology, Neurosurgery, and Psychiatry, 17, 235
- Ichikawa, H. (1970) - Proceedings of the 4th International Congress on Rural Medicine, Useda, Japan, 30th September to 4th October, 1967, 77
- Ikeda, Y. (1968) - Untitled Report on the Toxicity of Some Mercury Compounds, submitted to WHO
- Imura, N., Sukegawa, F., Pan, S., Nagao, K., Kim, J., Kwan, T., and Ukita, T. (1971) - Science, 172, 1248-9
- Innes, J.R.M., Ulland, V.M., Valerio, M.G., Petrucelli, L., Fishbein, L., Hart, E.R., Pallotta, A.J., Bates, R.R., Falk, H.L., Gart, J.J., Klein, M., Mitchell, I., and Peters, J. (1969) - Journal of the National Cancer Institute, 42, 1101
- Inoko, M. (1981) - Environmental Pollution (Series B), 2, 3-10

- Jalili, M.A. and Abbasi, A.H. (1961) - British Journal of Industrial Medicine, 18, 303

- Jensen, S. and Jernelov, A. (1967) - Nordforsk. Biocid. Inform., 10, 4

- Jensen, S. and Jernelov, A. (1968) - Nordforsk. Biocid. Inform., 14, 3

- Jensen, S. and Jernelov, A. (1969) - Nature, 223, 753-4

- Jernelov, H. (1969) - in Chemical Fallout, Charles C.Thomas, Springfield, Illinois, 75

- Joensüü, O.I. (1971) - Science, 172, 1027

- Jones, M.M. and Harbison, R.D. (1974) - Research Communications in Chemical Pathology and Pharmacology, 7, 389

- Jones, M.M., Pratt, T.H., Mitchell, W.G., Harbison, R.D., and MacDonald, S. (1976) - Journal of Inorganic and Nuclear Chemistry, 38, 613-6

- Kitamura, S. and Sumino, K. (1972) - Japanese Journal of Hygiene, 27, 123

- Kluge, H., Tschubel, H., and Zitek, A. (1938) - Zeitschrift für Unterslebensmittel, 76, 321

- Korring, P. and Hagel, P. (1973) - in Problems of the Contamination of Man and his Environment by Hg and Cd, Proceedings of an International Symposium, Luxemburg, 3rd-5th July, 1973, Commission of the European Communities, Luxemburg, 279

- Kudsk, F.N. (1965) - Acta Pharmacologia et Toxicologica, 23, 250

- Kurland, L. and Firo, L.N. (1960) - World Neurology, 1, 320

- Ladd, A.C., Goldwater, L.J., and Jacobs, M.B. (1964) - Archives of Environmental Health, 9, 43

- Landner, L. (1971) - Nature, 230, 452

- Larsson, J.E. (1970) - Swedish Environmental Protection Board Research Secretariat, Stockholm, Sweden

- Magos, L. (1967) - Environmental Research, 1, 323

- Magos, L. (1968) - British Journal of Industrial Medicine, 25, 315

- Magos, L. (1975) - British Medical Bulletin, 31, 241

- Magos, L. and Jawad, A.M. (1973) - Journal of Science of Food and Agriculture, 24, 1305

- Magos, L., Clarkson, T.W., and Greenwood, M.R. (1973) - Journal of Toxicology and Applied Pharmacology, 26, 180

- Magos, L., Sugata, Y., and Clarkson, T.W. (1974) - Journal of Toxicology and Applied Pharmacology, 28, 367

- McBirde, B.C. and Edwards, T.L. (1977) - in Biological Implications of Metals in the Environment, ERDA Symposium Series No. 42, 1-19

- Miller, V.L., Klavano, P.A., and Csonka, E. (1960) - Journal of Toxicology and Applied Pharmacology, 2, 344

- Mitra, S. and Ghosh Dastidar, P. (1976) - Science and Culture, 42, 443-51

- Nagase, H., Ose, Y., Sato, T., and Ishikawa, T. (1982) - Science of the Total Environment, 25, 133-42

- Norseth, T. and Clarkson, T.W. (1971) - Archives of Environmental Health, 22, 568

- Olson, B.H. and Cooper, R.C. (1976) - Water Research, 10, 113-6

- Ostlund, K. (1969) - Acta Pharmacologia et Toxicologica, Supplement, 1, 132

- Piechocka, J. (1968) - Rocznik Zaklad Higieny, 19, 385

- Piechocka, J. and Kranze, S. (1967) - Roczniki Pzh., 18, 711

- Rahola, T., Hattula, T., Korolainen, A., and Miettinen, J.K. (1971) - Annals of Clinical Research,

- Report of an International Committee (1969) - Archives of Environmental Health, 19, 891

- Rogers, R.D. (1976) - Journal of Environmental Quality, 5, 454-8

- Rogers, R.D. (1977) - US Environmental Protection Agency EPH-600/3-77-007

- Rupp, N.W. and Paffenberger, G.C. (1971) - Journal of the American Dental Association, 82, 401

- Sax, N.I. (1974) - in Industrial Pollution, Van Nostrand Reinhold, 215

- Smith, R.L. (1973) - The Excretory Function of the Bile, Chapters 3-8, Chapman and Hall, London

- Spence, R.W. (1971) - Journal of Medicine, 285, 971

- Stillings, B.R., Lagally, H., Bauersfeld, P., and Soares, (1974) - Journal of Toxicology and Applied Pharmacology, 30, 243

- Suzuki, T., Furukawa, K., and Tonomura, K. (1968) - Journal of Fermentation Technology, 46, 1048

- Swedish Expert Group (1971) - Nordisk Hygienisk Tidskrift Supplement No. 4
- Swensson, A. (1952) - Acta Medica Scandinavica, 143, 365
- Swensson, A. and Ulfvarson, U. (1968) - Acta Pharmacologia et Toxicologica, 25, 259, and Acta Pharmacologia et Toxicologica, 26, 273
- Takakashi, H. and Hirayama, K. (1971) - Nature, 232, 210
- Takeuchi, T. (1970) - International Conference on Environmental Mercury Contamination, Ann Arbor, Michigan, 30th September to 2nd October, 1970
- Task Group on Metal Accumulation (1973) - Environ. Physiol. Biochem., 3, 65
- Tejning, S. (1967) - Report 670210 from Clinic of Occupational Medicine in Lasarettet at Lund (in Swedish)
- Tonomura, K., Kanazaki, F. (1969a) - Hakko Kogaku Zasshi, 47, 430
- Tonomura, K., Kanazaki, F. (1969b) - Biochimica et Biophysica Acta, 184, 227
- Ulfvarson, U. (1962) - Intern. Ach. Gewerbepathol. Gewerbehyg., 19, 412-22
- US Minerals Yearbook (1968), 1-2, 377
- Vallee, B.L. and Ulmer, D.D. (1972) - Annual Review of Biochemistry, 41, 91
- Vaughn, W.W. (1967) - US Geological Survey Circular 540, 1-8, US Government Printing Office, Washington, DC
- von Massman, W. (1957) - Zentralblatt Arbeitsmedizine Arbschutz, 7, 9
- Vostal, J.J. and Clarkson, T.W. (1973) - Journal of Occupational Medicine, 15, 649
- Wein, R. (1939) - Quarterly Journal of Pharmacy, 12, 212
- Westöö, G. (1966) - Acta Chimica Scandinavica, 20, 2131-7
- Westöö, G. (1967) - Var Föda, 19, 121
- Westöö, G. (1968) - Acta Chimica Scandinavica, 22, 2277
- Westöö, G. (1969) - Var Föda, 21, 17
- Wood, J.M., Kennedy, F.S., and Rosen, C.G. (1968) - Nature, 220, 173
- Wood, J.M., Penley, M.W., and De Simone, R.E. (1971) - Mercury Handbook, International Atomic Energy Commission
- Yamaguchi, S., Matsumoto, H., Hoshida, M., Matsuo, S., and Kaku, S. (1971) - Archives of Environmental Health, 23, 196

- Zeyer, H.G. (1952) - Zentralblatt Arbeitsmedizine Arbschutz, 2, 68

CHAPTER IX
MAJOR CASE STUDIES

INTRODUCTION

Accidental and intentional mercury poisonings have a long history but, before the 1950's, there are no recorded cases of environmental mercury problems. Such problems were highlighted in 1950 when a factory in Minamata, Japan, which discharged mercury-containing waste into Minamata Bay, caused disaster. The bay is quite small and is surrounded by a population of about 10000 who live in small fishing villages. The first case of Minamata Disease (since diagnosed to be methylmercury poisoning) occurred in 1953 and was followed by many other cases. The epidemic of poisoning did not come to the attention of public health authorities until 1956, by which time there were about 80 cases. In late 1958, the pathological symptoms of Minamata Disease were identified with those found in accidental alkylmercury poisoning and, in 1959, the causative agent was shown to be methylmercury present in the fish and shellfish of the bay. Neither governmental agencies nor the factory management aided the Japanese study group in identifying the causative agent or its source. The resulting difficulties delayed corrective action so much so that, in 1964/5, a second outbreak of the disease occurred, this time among fish-eaters living near Niigata, Japan. The Minamata poisonings involved 121 cases with 46 deaths. The Niigata poisoning involved 47 cases with 6 deaths.

The scientific and epidemiological studies of the Minamata (1953-60) and Niigata (1965) disasters are widely known and the disease has been identified as being neuropathia arising from the intake of large quantities of fish which accumulated methylmercury in their flesh. A study group on the "strange Minamata disease" was first organised by the medical group of Kumamoto University in 1956, and was followed by many other such studies.

The details of the Minamata and Niigata cases are widely known to-day and various aspects have been discussed by Harada (1968), Kitamura (1968), Irukayama (1977), Takizawa (1977), Tsubaki (1977), and many others. For this reason, the details will not be presented in this monograph.

Mercury poisoning cases in human beings, as reported from Iraq and New Mexico, and some industrial exposure cases are discussed below, followed by cases of poisoning of domestic animals and wildlife in various countries.

Table 9.1 Details of the Epidemic of Methylmercury Poisoning in Iraq (Bakir et al, 1973)

Province	Crop	Amount Ingested* (tonne [%]**)	P	A	D
Nineveh	wheat	31766 [46]			
	barley	9102 [41]	584000	592	9
Kirkuk	wheat	10659 [36]			
	barley	4392 [40]	242000	814	42
Arbil	wheat	10105 [46]			
	barley	1538 [38]	222000	242	10
Baghdad	wheat	3783 [26]			
	barley	377 [8]	449000	107	34
Qadissiya	wheat	3363 [51]			
	barley	715 [24]	269000	1345	73
Waset	wheat	2741 [16]			
	barley	511 [4]	231000	688	11
Muthanna	wheat	1789 [47]			
	barley	1665 [84]	92000	701	79
Diala	wheat	1627 [31]			
	barley	962 [21]	262000	675	12
Thi-Qar	wheat	1456 [21]			
	barley	382 [5]	366000	191	30
Sulaimaniya	wheat	1453 [10]			

continued

Table 9.1 (continued) Details of the Epidemic of Methylmercury Poisoning in Iraq (Bakir et al, 1973)

Province	Crop	Amount Ingested* (tonne [%]**)	P	A	D
Sulaimaniya	barley	42 [1]	272000	4	1
Babylon	wheat	1350 [21]			
	barley	1095 [20]	284000	1083	128
Dhok	wheat	1153 [38]			
	barley	523 [69]	84000	101	1
Anbar	wheat	831 [16]			
	barley	224 [18]	184000	16	5
Maysan	wheat	693 [24]			
	barley	699 [22]	241000	106	8
Kerbala	wheat	276 [56]			
	barley	25 [7]	89000	141	5
Basra	wheat	155 [24]			
	barley	10 [19]	27000	7	2
TOTAL	wheat	73201			
	barley	22262	3500000	6000	500

* seed grain, ** percentage of total seeds planted, P: rural population, A: victims admitted to hospital, D: victims dying in hospital

METHYLMERCURY POISONING IN IRAQ

The epidemic in Iraq was caused by exposure to methylmercury when people ate homemade bread prepared from wheat seed which had been treated with a methylmercurial fungicide. The treated grain was distributed to all of the provinces of the country, with over 50% of the total going to the provinces of Ninevah, Kirkuk, and Arbil. Consumption of bread made from the treated grain started as early as October in some families. Poisoning cases were hospitalised from early in January, 1972. The details of the epidemic of methylmercury poisoning in Iraq are shown, province by province, in table 9.1. It was confirmed that methylmercury was the real cause of poisoning. Its content varied from 3.7 to 14.9 μg Hg/g of wheat. The mean methylmercury content of wheat flour samples was 9.1 μg, with a range of 4.8 to 14.6 μg. The severity of the signs and symptoms was dose-dependent and ranged from ataxia (slight unsteadiness of gait) to such gross discoordination that the patient could not walk. Defects in vision including blurred vision and constriction of the visual field, leading to blindness in severe cases, were also observed. Hearing difficulties, together with slurred speech, were exhibited by patients with high mercury concentrations in their blood. Fatalities resulted from an apparent failure of the central nervous system (table 9.2).

Although mercury can enter the human body as a result of oral ingestion, inhalation, or skin contact, the predominant cause of Hg poisoning in the epidemic appeared to be the consumption of homemade bread prepared from treated grain. Males and females of all ages were affected, from infants under 1 year of age to elderly people. Table 9.3 indicates the distribution of cases according to age and sex.

Table 9.2 Frequency of Symptoms of Methylmercury Poisoning According to the Concentration of Mercury in the Blood and the Estimated Total Amount of Mercury Ingested (Bakir et al, 1973)

C_B(ng/ml)	P_i(days)	P_l(days)	t_m	Symptom	N_v	C_s(%)
0 - 100	43	-	-	paresthesia	21	9.5
				ataxia		5
				visual changes		0
				dysarthria		5
				hearing defects		0
				death		0
101-500	43	-	-	paresthesia	19	5
				ataxia		0
				visual changes		0
				dysarthria		5
				hearing defects		0

continued

Table 9.2 (continued) Frequency of Symptoms of Methylmercury Poisoning According to the Concentration of Mercury in the Blood and the Estimated Total Amount of Mercury Ingested (Bakir et al, 1973)

C_B(ng/ml)	P_i(days)	P_l(days)	t_m	Symptom	N_v	C_s(%)
101-500	43	-	-	death		0
501-1000	43	16	115	paresthesia	19	42
				ataxia		11
				visual changes		21
				dysarthria		5
				hearing defects		5
				death		0
1001-2000	41	18	96	paresthesia	17	60
				ataxia		47
				visual changes		53
				dysarthria		24
				hearing defects		0
				death		0
2001-3000	55	26	33	paresthesia	25	79
				ataxia		60
				visual changes		56
				dysarthria		25
				hearing defects		12.5
				death		0
3001-4000	58	32	30	paresthesia	17	82
				ataxia		100
				visual changes		58
				dysarthria		75
				hearing defects		36
				death		17
4001-5000	68	38	20	paresthesia	7	100
				ataxia		100

continued

Table 9.2 (continued) Frequency of Symptoms of Methylmercury Poisoning According to the Concentration of Mercury in the Blood and the Estimated Total Amount of Mercury Ingested (Bakir et al, 1973)

C_B(ng/ml)	P_i(days)	P_l(days)	t_m	Symptom	N_v	C_s(%)
4001-5000	68	38	20	visual changes		83
				dysarthria		85
				hearing defects		66
				death		28

C_B: concentration of Hg in blood, P_i: mean period of ingestion, P_l: mean latent period, t_m: mean time at which sample was obtained, N_v: number of victims, C_s: percentage of victims suffering from given symptom

Most of the mercury in whole blood is in organic form and is located in the red blood cells. Plasma, milk, urine, cerebrospinal fluid, and amniotic fluid differed from whole blood in that the concentration of total mercury was much lower and the proportion of inorganic mercury was higher (table 9.4).

Table 9.3 Distribution of Cases of Poisoning Admitted to Hospital, and of Deaths Caused by Poisoning, According to Age and Sex

Age (years)	Victim Type	Number of Victims	Fatalities	Fraction(%)
1	male	8	0	0
1	female	2	0	0
1	total	10	0	0
1 - 4	male	383	33	8.6
1 - 4	female	378	36	9.5
1 - 4	total	761	69	9.1
5 - 9	male	736	63	8.6
5 - 9	female	734	58	7.9

continued

Table 9.3 (continued) Distribution of Cases of Poisoning Admitted to Hospital, and of Deaths Caused by Poisoning, According to Age and Sex

Age (years)	Victim Type	Number of Victims	Fatalities	Fraction(%)
5 - 9	total	1470	121	8.2
10 - 19	male	606	45	7.4
10 - 19	female	650	55	8.5
10 - 19	total	1256	100	8.0
20 - 29	male	380	18	4.7
20 - 29	female	520	43	8.3
20 - 29	pregnant female	17	12	70.5
20 - 29	total	917	73	8.0
30 - 39	male	416	15	3.6
30 - 39	female	483	25	5.2
30 - 39	pregnant female	12	2	16.7
30 - 39	total	913	42	4.6
40 - 49	male	240	7	2.9
40 - 49	female	268	17	6.3
40 - 49	pregnant female	2	0	0
40 - 49	total	510	24	4.7
50 - 59	male	164	13	7.9
50 - 59	female	146	12	8.2
50 - 59	total	310	25	8.1
60 +	male	211	3	1.4
60 +	female	172	2	1.2
60 +	total	383	5	1.3
ALL AGES	male	3144	197	6.3
	female	3353	248	7.4
	pregnant female	31	14	45.0
	total	6530	459	7.0

Treatment was carried out by the administration of several mercury-binding compounds (D-penicillamine, N-acetyl-DL-penicillamine, thiol resin) to patients in order to promote the excretion of mercury.

The reader is referred to an excellent article by Bakir et al (1973) for a full description of the methylmercury poisoning in Iraq.

Table 9.4 Total Amounts of Mercury in Samples of Biological Fluid, Compared with Heparinised Whole Blood Obtained from Patients with Methylmercury Poisoning (Bakir et al, 1973).

Fluid	N	Total Hg*	Correlation with blood	Inorganic Hg**
blood	224	100	1.0	7
plasma	14	18	0.8	22
milk	44	5	0.9	39
urine	21	6	0.1	73
cerebrospinal	5	6	-	-
amniotic	1	2	-	-

* percentage of whole blood, ** percentage of total

MERCURIAL FUNGICIDE POISONING IN NEW MEXICO

In August, 1969, a farmer (Mr H) and five of his neighbours in the area of Alamogordo, New Mexico, obtained waste seed grain from a local granary. The grain had been treated with organic mercurial fungicides. The six farmers used this grain as food for hogs. The father of one family (Mr H) began feeding his pigs with the grain in late August or early September. After some 2 to 3 weeks, a hog which had been fed 60% more grain than the others was slaughtered and the family ate the meat over the next 3 to 5 months. By mid-October, 14 of the other pigs had developed blindness, lack of coordination, and posterior paralysis. In the next 3 weeks, 12 of the 14 pigs died. Gait disturbances in the surviving pigs improved, but they remained blind.

In early December, one child in Mr H's family fell sick and, by late December, two other members of the family had developed the same illness. Mercury poisoning caused by the ingestion of contaminated pork was suspected and the patients were treated for mercury poisoning. At this time, the mother of the children was pregnant. Three months later, the baby was born.

Table 9.5 Concentration of Mercury (AAS analysis) in Hog Tissue and in Seed Grain Fed to the Given Hog (Curley et al, 1971)

Owner	Hog Tissue	Hg Content (ppm)	Grain (ppm Hg)
Mr H (slaughtered hog)	muscle	29.4	-
Mr H (slaughtered hog)	muscle	27.5	32.8
Mr H (blind hog)	brain	36.1	32.8
Mr H (blind hog)	liver	25.2	32.8
Mr H (blind hog)	kidneys	12.0	32.8
Mr H (blind hog)	muscle	14.2	32.8
Neighbour 1	brain	13.1	2.97
Neighbour 1	liver	17.3	2.97
Neighbour 1	kidneys	12.1	2.97
Neighbour 1	muscle	9.5	2.97
Neighbour 1	colon	24.2	2.97
Neighbour 2	brain	15.8	1.27
Neighbour 2	liver	21.0	1.27
Neighbour 2	kidneys	25.2	1.27
Neighbour 2	muscle	23.1	1.27
Neighbour 2	colon	6.5	1.27
Neighbour 3	brain	12.6	2.76
Neighbour 3	liver	3.5	2.76
Neighbour 3	kidneys	8.4	2.76
Neighbour 3	muscle	16.8	2.76
Neighbour 3	colon	14.7	2.76
Neighbour 4	brain	3.5	-
Neighbour 4	liver	8.8	-
Neighbour 4	kidneys	20.5	-
Neighbour 4	muscle	16.8	-
Neighbour 4	colon	7.9	-

continued

Table 9.5 (continued) Concentration of Mercury (AAS analysis) in Hog Tissue and in Seed Grain Fed to the Given Hog (Curley et al, 1971)

Owner	Hog Tissue	Hg Content (ppm)	Grain (ppm Hg)
Neighbour 5	brain	21.0	2.54
Neighbour 5	liver	10.5	2.54
Neighbour 5	kidneys	21.0	2.54
Neighbour 5	muscle	12.6	2.54
Neighbour 5	colon	24.2	2.54

Sample	Range (ppm)	Mean (ppm)	Standard Error (ppm)
brain	<3.5 - 21	12.5	± 3.46
liver	<3.5 - 21	12.2	± 3.05
kidneys	8.4 - 25.2	17.4	± 3.10
muscle	9.5 - 23.1	15.8	± 2.29
colon	6.5 - 24.2	15.5	± 3.81
grain	1.27 - 2.97	2.39	± 0.38

The hog, fed on large quantities of the waste grain and slaughtered for family consumption, contained high concentrations of Hg in its tissues, as did another sick and ultimately blind hog from the same herd. The waste grain fed to these animals contained 32.8ppm of mercury. Hogs belonging to the neighbours, and fed with grain from the same source, contained approximately the same concentration of mercury in their tissues.

The concentrations of mercury in serum, urine, and cerebrospinal fluid were determined in samples taken from the human victims. Urine samples obtained from Mr H, his son (13), and his two daughters (8 & 20) on the 8th January contained 0.16, 0.21, 0.20, and 0.06ppm of Hg, respectively. The concentrations of Hg in urine samples taken from the neighbours varied from less than 0.05 to 0.18ppm (33 samples). The content of Hg in serum samples from the neighbours averaged less than 0.02ppm (38 samples). The urine of Mrs H, who was pregnant at the onset of the children's illness, contained 0.09ppm of mercury on the 8th January and 0.18ppm on the 11th February.

Her serum contained 2.9ppm of Hg on the 22nd January and 0.47ppm on the 11th February. The amniotic fluid contained less than 0.02ppm of Hg on the 11th February. The concentration of Hg in the newborn baby's urine ranged from 2.70ppm at delivery, to 1.56ppm several days later.

These data clearly show that mercury accumulates in animal tissues and human body fluids, and confirm that compounds containing organic mercury were, in fact, the causative agents in the poisoning incident. The change in the Hg concentration in the serum and urine of the mother after delivery, and the contents of mercury in the urine of the newborn baby indicated placental transfer to the foetus.

EXPOSURE OF MERCURY WORKERS

El-Sadik et al (1970) have conducted a study of the exposure of workers to mercury at a NaOH producing plant. This plant electrolyses NaCl in order to obtain NaOH, using Hg in the electrolytic cells.

Table 9.6 Clinical Findings Among Various Age Groups for Subjects Exposed to Mercury for Less Than Six Months (El-Sadik et al, 1970)

Age	N_w	Hg_u (μg/l)	Hg_s (μg)	Clinical Sign	N_c	P_c (%)
25-29	1	96	7.6	any	1	100
				pastomatitis	0	0
				stomatitis	0	0
				low blood pressure	1	100
				tremors	1	100
				behavioural changes	1	100
				deep reflex	0	0
30-39	9	48	6.6	any	7	77.7
				pastomatitis	2	22.2
				stomatitis	4	44.4
				low blood pressure	1	11.1
				normal blood pressure	6	66.6
				hypertension	2	22.2

continued

Table 9.6 (continued) Clinical Findings Among Various Age Groups for Subjects Exposed to Mercury for Less Than 6 Months (El-Sadik et al, 1970)

Age	N_w	Hg_u(µg/l)	Hg_s(µg)	Clinical Sign	N_c	P_c(%)
30-39	9	48	6.6	tremors	1	11.1
				behavioural changes	2	22.2
				deep reflex	1	11.1
40-49	3	132	3.7	any	3	100
				pastomatitis	2	66.6
				stomatitis	2	66.6
				low blood pressure	1	33.3
				normal blood pressure	1	33.3
				hypertension	1	33.3
				tremors	1	33.3
				behavioural changes	0	0
				deep reflex	1	33.3
Total	13	71	6.0	any	11	84.6
				pastomatitis	4	30.8
				stomatitis	6	46.2
				low blood pressure	3	23.1
				normal blood pressure	7	53.9
				hypertension	1	23.1
				tremors	3	23.1
				behavioural changes	3	23.1
				deep reflex	2	15.4

N_w: number of workers, Hg_u: average mercury in urine, Hg_s: average mercury in saliva, N_c: number of cases with given clinical sign, P_c: percentage of cases with given clinical sign

Method: Each worker exposed to Hg in the plant was examined. The examination included the subject's personal and occupational history. The mouth, skin, heart, chest, abdomen, and central nervous system were examined. During work, a urine sample was taken from each individual in order to determine the mercury level. Saliva was collected from each subject by asking him to spit all of the saliva in his mouth into a Pyrex container. Of the 13 workers exposed to Hg for less than six months, 11 had complaints, and nine of these involved neurasthenia (table 9.6). Similarly, of 13 workers with exposure times ranging from 6 months to 3 years, 11 reported complaints, and 10 of these were neurasthenic (table 9.7).

Table 9.8 shows that 26 workers had exposures of three years or more, with only one worker not having complaints, and 21 complaining of neurasthenia. All cases of tremor noted among these workers affected the hands. Hypertension was noted only in workers over 30 years old.

Table 9.7 Clinical Findings Among Various Age Groups for Subjects Exposed to Mercury for Between Six Months and Three Years (El-Sadik et al, 1970)

Age	N_w	Hg_u (μg/l)	Hg_s (μg)	Clinical Sign	N_c	P_c (%)
20-24	2	184	4.1	any	2	100
				pastomatitis	1	50
				stomatitis	2	100
				normal blood pressure	2	100
				tremors	1	50
				behavioural changes	1	50
				deep reflex	0	0
30-39	10	112	3.6	any	8	80
				pastomatitis	6	60
				stomatitis	7	70
				low blood pressure	2	20
				normal blood pressure	4	40
				hypertension	4	40
				tremors	5	50
				behavioural changes	5	50
				deep reflex	2	20
40-49	1	4	8.6	any	1	100

continued

Table 9.7 (continued) Clinical Findings Among Various Age Groups for Subjects Exposed to Mercury for Between Six Months and Three Years (El-Sadik et al, 1970)

Age	N_w	Hg_u(μg/l)	Hg_s(μg)	Clinical Sign	N_c	P_c(%)
40-49	1	4	8.6	pastomatitis	0	0
				stomatitis	1	100
				normal blood pressure	1	100
				tremors	0	0
				behavioural changes	1	100
				deep reflex	0	0
Total	13	101	4.1	any	11	84.6
				pastomatitis	7	53.9
				stomatitis	10	76.9
				low blood pressure	2	15.4
				normal blood pressure	7	53.9
				hypertension	4	30.8
				tremors	6	46.2
				behavioural changes	7	53.9
				deep reflex	2	15.4

N_w: number of workers, Hg_u: average mercury in urine, Hg_s: average mercury in saliva, N_c: number of cases with given clinical sign, P_c: percentage of cases with given clinical sign

DOMESTIC ANIMALS AND WILDLIFE

A few reports are available on the incidence of mercury poisoning in livestock. In Sweden, where certain bird populations decreased drastically, investigations revealed the presence of large amounts of mercury in the dead birds. The death of the birds was related to the extensive use of methylmercury dicyanodiamide as a fungicide in Swedish agriculture. Most of a herd of 44 adult swine and 5 litters of pigs, which were fed for several weeks on a diet of which one half consisted of grain treated with a

methylmercuric compound, died within a three month period (Kahrs, 1968).
In June, 1970, a resident of the Washington, DC area reported that his cat was showing symptoms of mercury poisoning. His description was almost identical to the behaviour of poisoned cats in Minamata, i.e. unsteadiness, frequent falls, circling movements, and convulsions. The cat died and kidney and liver tissue samples were analysed for mercury. The mercury residues observed in the liver and kidney samples were 63ppm and 7ppm, respectively. The reports of Takeuchi (1970) indicate that cats affected by methylmercury poisoning in Minamata had mercury levels in liver samples which ranged from 37 to 145.5ppm for 15 cats, while the mercury levels in kidney samples ranged from 12.2 to 36.1ppm for 7 cats. Those cats which were not affected had a level of 0.6 to 6.6ppm in the liver.

Table 9.8 Clinical Findings Among Various Age Groups for Subjects Exposed to Mercury for More than Three Years (El-Sadik et al, 1970)

Age	N_w	Hg_u (μg/l)	Hg_s (μg)	Clinical Sign	N_c	P_c(%)
25-29	5	66	5.0	any	5	100
				pastomatitis	4	80
				stomatitis	4	80
				low blood pressure	2	40
				normal blood pressure	3	60
				hypertension	0	0
				tremors	5	100
				behavioural changes	5	100
				deep reflex	3	60
30-39	17	64	5.8	any	16	94.1
				pastomatitis	4	23.5
				stomatitis	14	82.4
				low blood pressure	5	29.4
				normal blood pressure	8	47
				hypertension	4	23.5
				tremors	14	82.4

continued

Table 9.8 (continued) Clinical Findings Among Various Age Groups for Subjects Exposed to Mercury for More than 3 Years (El-Sadik et al, 1970)

Age	N_w	$Hg_u(\mu g/l)$	$Hg_s(\mu g)$	Clinical Sign	N_c	$P_c(\%)$
30-39	17	64	5.8	behavioural changes	12	70.6
				deep reflex	9	52.9
40-49	4	39	6.4	any	4	100
				pastomatitis	3	75
				stomatitis	4	100
				normal blood pressure	1	25
				hypertension	3	75
				tremors	4	100
				behavioural changes	3	75
				deep reflex	2	50
Total	26	61	5.7	any	25	96.2
				pastomatitis	11	42.3
				stomatitis	22	84.6
				low blood pressure	7	26.9
				normal blood pressure	12	46.2
				hypertension	7	26.9
				tremors	23	88.5
				behavioural changes	20	76.9
				deep reflex	14	53.8

N_W: number of workers, Hg_U: average mercury in urine, Hg_S: average mercury in saliva, N_C: number of cases with given clinical sign, P_C: percentage of cases with given clinical sign

According to the reports of the United States Fish and Wildlife Service (1970), two bald eagles which were found dead in Minnesota during 1970 had lethal amounts of mercury in their kidneys. The kidneys contained mercury concentrations of 130 and 117ppm. It was concluded, by the Study Group on Mercury Hazards (1970), that liver/kidney residues of 30ppm in birds indicate a critical exposure. It is believed that the eagles had accumulated mercury via the ingestion of fish (Fish and Wildlife Service, 1970).

FISH

Fish are generally unaffected by high mercury levels, although a few incidences of death are known. In Boone Reservoir, Tennessee, over 500000 fish were killed between the 9th and 13th July, 1968, while over 2300000 fish were killed between the 9th and 13th May, 1969 by phenylmercuric acetate (FWPCA, 1969).

In fish, mercury is present mainly in organic form. Older and larger species of fish contain higher concentrations of mercury than do younger and smaller ones. Thus, large tuna or swordfish are found to contain between 0.5 and 1.5μg/g dry weight (Miller et al, 1972) while small myctophids contain 0.1 to 0.3μg/g dry weight (Gibbs et al, 1974).

Except for a few exceptions, such as Pacific blue marlin (Shultz et al, 1976), all fish seem to contain mercury predominantly in the monomethyl form. In most cases, the concentration is higher in the liver and spleen than in the muscles. The accumulation rate of organic mercury is higher than that of inorganic compounds. The biological half-life of organic mercury is also longer than that of the inorganic form.

The transfer time of mercury from water to fish was investigated by Klaverkamp et al (1983), using gamma-emitting isotopes such as ^{203}Hg, ^{59}Fe, ^{60}Co, ^{65}Zn, ^{75}Se, and ^{134}Cs in the Experimental Lake Area of North-Western Ontario, Canada. The results revealed a considerable time lag between contamination and the transfer of contaminant to the fish species.

MERCURY AND SELENIUM IN MARINE MAMMALS/FISH/BIRDS

The existence of a protective mechanism against the toxic effects of mercury was suggested by the simultaneous uptake (sometimes in equimolar amounts) of selenium and mercury by certain marine mammals and birds (Koeman et al, 1975) and fish (Ganther et al, 1972; Mackay et al, 1975). This has been confirmed by laboratory studies of birds and mammals (Chen et al, 1975; El-Begearmi et al, 1977; Friedman et al, 1978), but the limited information on fish is insufficiently clear (Martoja & Martoja, 1978) or is conflicting (Freeman et al, 1978; Cappon & Smith, 1981). Stimulation of Hg uptake by moderate selenium exposure occurs in goldfish (Heisinger et al, 1979).

Studies of trace elements in marine organisms from the north-west Mediterranean (northern Tyrrhenian Sea), which revealed widespread high mercury levels, prompted Leonzio et al (1982) to study the relationship between mercury and selenium in fish muscle, using the striped mullet (***Mullus barbatus L.***) as a test species. This mullet has a broad geographical distribution and has a good food market. They found that, where the levels of Hg were high, those of selenium were low, and vice versa. They plotted the sum of the concentration of these two elements against the age of the striped mullet and obtained a single function with a highly significant linear correlation. These findings suggest that, in the muscles of these animals, there are selenium receptors, whose number increases during their life-time. It is likely that these receptors can be occupied by mercury in proportion to its bioavailability in the environment. The competition between Hg and Se for a given receptor may also explain their amply demonstrated toxicological antagonism.

OTHER AQUATIC ORGANISMS

Elevated levels of Hg and Pb are found in the sediments of the oyster fishery at the seaward limit near to the Tamar Bridge at Lynher (Bland et al, 1982). There are no significant sources of Hg in the catchment area (Perkins, 1971) and this suggests that the high concentrations at the site and in the Tamar river are associated with sewage inputs, as in the case of the Plym Estuary (Millard & Herbert, 1981). However, analyses of oysters did not reveal such high levels of mercury as might have been expected. (The possibility of loss of some of the mercury compounds during cleaning prior to chemical analysis could be ruled out). The reported values in the oysters and sediments are shown in table 9.9

AQUATIC BIOAVAILABILITY

Upon discharge into an aquatic environment, mercury, like other metals, is partitioned between the solid and liquid phases, wherein speciation occurs depending upon the ligands bonding it. Mercury exchanges slowly, compared to ferrous and non-ferrous base metals. An increased pH of the sediment appears to increase the availability of Hg to marsh plants, perhaps by stimulating the conversion of Hg^{2+} to Hg° (Gambrell et al, 1977). Biological processes typically double in rate for every 10°C increment in temperature (a Q_{10} of 2). Such a value has been observed in the muscle tissue of blue-gill exposed to methylmercury (Cumber et al, 1978), which exchanges very slowly in fish tissues.

Carriers may exist which are selective for specific metal complexes. Carrier-facilitated processes are not the only mode of metal uptake. Forms of the metal which are lipid-soluble, or of reduced polarity, could cross membranes via diffusion. A reduced polarity is the most likely explanation for the exceptionally high bioavailability of ethylmercury (Philips & Buhler, 1978). It is noted that Hg° is more available, as compared to Hg^{2+}, because of the lipid solubility of Hg° (Hughes, 1957).

Table 9.9 Mercury Contents of Oysters from the Lynher Estuary, UK

Location	Hg (ppm, dry weight)	References
	Sediments	
Lynher	2.1	Bland et al, 1982
Tamar	0.2 - 1.5	Ackroyd & Marsh, 1981
	Oysters (***O. edulis***)	
Lynher, UK	0.17 - 0.31	Bland et al, 1982
	Oysters (***C. gigas***)	
Helford, UK	0.76	Thornton et al, 1975
	Oysters (***C. virginia***)	
San Antonio Bay, USA	0.05	Sims & Presley, 1976

A higher concentration of mercury is found in predatory fish. Biomagnification through successive trophic levels would be expected if rigorous digestion were the cause of such concentrations. Perhaps the transformation of metals by bacteria in the gut of some species may be involved where high concentrations coincide with a high trophic level.

The tissue distribution of metals varies according to the source of the uptake. Differences in tissue distribution between animals exposed to metals in food and those exposed to metal-enriched solutions have been found for mercury in oysters (Cunningham & Trip, 1975), mussels (Pentreath, 1973), and fish (Pentreath, 1976a,b,c). Pentreath's experimental work on fish showed conclusively that food, rather than the solution, was the major source of metal in fish as well as in other aquatic organisms. Mercury in the methylated form is several times more available to biota than is the inorganic form (Kramer & Neidhart, 1975).

Several studies which demonstrated that fish accumulate Hg (Jernelov, 1970; Gillispie, 1972; Bongers & Khattak, 1972) and other metals (DeLisle, 1975) in a closed sediment/water system, did not identify the vector of uptake (Luoma, 1983). Kudo and Mortimer (1979) also found much higher concentrations of Hg in "guppies" which were held in contact with labelled sediments, than in fish which were physically separated from the sediment by nylon mesh.

Phillips and Gregory (1979) observed that the efficiency of methylmercury uptake from food by pike ranged from 58 to 89% in studies where the metal was added to synthetic foods, tissue homogenates, or fed to prey for short periods. However, when the pike obtained their methylmercury from prey contaminated over their lifetime in nature, the assimilation efficiencies were 20% or less. Longer exposure to the metal resulted in proportionally greater quantities of Hg in slowly exchanging, and (presumably) more strongly bound pools of Hg in the organisms (Luoma, 1977).

MERCURY POLLUTION OF THE LAKE ERIE ECOSPHERE

When significantly high levels of mercury were found in certain fish species of the Great Lakes, it became apparent that Lake Erie was polluted with mercury and therefore studies were initiated (Pillay et al, 1972) in order to investigate the situation. Samples of air, lake sediment, plankton/algae, fish, etc, were collected at various sampling points. Brain tissue from subjects who had lived in the area for several years was collected from neighbouring hospitals and analysed for mercury content.

MERCURY IN AIR PARTICULATES

Assays revealed that the mercury contents of air particulates were greater than 0.1μg, although there were large variations due to differences in the atmospheric conditions, which determine the residence time of particulates in air.

MERCURY IN SEDIMENTS AND PLANKTON/ALGAE

The mercury contents of sediments and plankton/algae were determined. The results revealed higher levels of mercury in the sediments and plankton/algae from the southern shore of the Western Basin of Lake Erie. There were generally increased levels of mercury in the plankton/algae from sampling points when the sediment levels were high.

MERCURY IN FISH

Fish composites prepared from the edible tissues of eleven different species were analysed for mercury and the results are given in table 9.10.

Table 9.10 Mercury Content of the Edible Tissues of Lake Erie Fish

Species	Basin	Mercury Content (μg/g of raw tissue)*
Walleye	western	0.79 (25)**
Walleye	central	0.65 (25)
Walleye	eastern	0.33 (25)
Yellow Perch	western	0.61 (25)
Yellow Perch	central	0.49 (25)
Yellow Perch	eastern	0.29 (25)
White Bass	western	0.60 (25)
White Bass	central	0.72 (25)
White Bass	eastern	0.43 (25)
Channel Catfish	western	0.36 (25)
Channel Catfish	central	0.42 (20)
Freshwater Drum	western	0.67 (25)
Freshwater Drum	central	0.62 (20)
Freshwater Drum	eastern	0.30 (25)
Carp	western	0.23 (25)
Carp	central	0.35 (17)
Carp	eastern	0.36 (14)
Coho Salmon	western	0.69 (20)
Coho Salmon	central	0.58 (10)
Coho Salmon	eastern	0.51 (13)
White Sucker	western	0.55 (24)

continued

Table 9.10 (continued) Mercury Content of the Edible Tissues of Lake Erie Fish

Species	Basin	Mercury Content (μg/g of raw tissue)*
White Sucker	central	0.56 (8)
White Sucker	eastern	0.35 (25)
Cizzard Shad	western	0.20 (25)
Cizzard Shad	central	0.21 (15)
Cizzard Shad	eastern	0.26 (18)
Smallmouth Bass	central	0.55 (14)
Smelt**	eastern	0.30 (10)

* composite of edible tissues, ** the numbers in parentheses refer to the number of samples of fish from a particular species used in preparing the composite, *** mercury content of the whole fish

MERCURY IN HUMAN BRAIN TISSUE

Measured levels in brain tissue ranged from a low of 0.02 μg/g of raw tissue to a high of 2.27 μg/g.

Pillay et al (1972) have concluded that Hg release due to mining and metallurgical operations was non-existent in the Erie Basin. The mercury levels in the coal mined in the Lake Erie region did not exhibit unusually high values, indicating that the amount of mercury transferred from the pedosphere to the hydrosphere was minimal. Therefore, it has been assumed that the elevated levels of mercury in the fauna and flora of Lake Erie are due to the intentional use of mercury in that region.

REFERENCES

- Ackroyd, D.R. and Marsh, J.G. (1981) - referred to by Bland et al (1982)
- Bakir, F., Damluji, S.F., Amin-Zaki, L., Murtadha, M., Khalidi, A., Al-Rawi, N.Y., Tikriti, S., Dhahir, H.I., Clarkson, T.W., Smith, J.C., and Doherty, R.A. (1973) - Science, 181, 230

- Bland, S., Ackroyd, D.R., Marsh, J.G., and Millward, G.E. (1982) - Science of the Total Environment, 22, 235-41

- Bongers, L.H. and Khattak, M.N. (1972) - USEPA Publication No. 16080 HVA, 245

- Bryan, G.W. (1976) - in Marine Pollution by R.Johnston (editor), Academic Press, London, p185

- Cappon, C.J. and Smith, J.C. (1981) - Archives of Environmental Contamination and Toxicology, 10, 305-19

- Chen, R.W., Lacy, V.L., and Whanger, P.D. (1975) - Research Communications in Chemical Pathology and Pharmacology, 12, 297-308

- Cumber, H., Curtis, E.H., and Blaylock, B.G. (1978) - Environmental Pollution, 17, 311-21

- Cunningham, P.A. and Trip, M.R. (1975) - Marine Biology, 31, 321-34

- Curley, A., Vincent, A.S., Edward, F.G., Robert, E.H., Barthel, W.F., Likosky, W.H., and Paul, E.P. (1971) - Science, 172, 65

- DeLisle, C.E. (1975) - International Conference on Heavy Metals in the Environment, Toronto, Canada, 821

- El-Begearmi, M.M., Sunde, M.L., and Ganther, H.E. (1977) - Poultry Science, 56, 313-22

- El-Sadik, Y.M. and El-Dakhakhny, A.A. (1970) - American Industrial Hygiene Association Journal, 31, 705

- Fish and Wildlife Service (1970) - Bureau of Sport, Fisheries, and Wildlife, Department of the Interior, News Release, 2nd July, 1970

- Freeman, H.C., Shum, G., and Uthe, J.F. (1978) - Journal of Environmental Science and Health, A13, 235-40

- Friedman, M.A., Eaton, L.R., Carter, W.H. (1978) - Bulletin of Environmental Contamination and Toxicology, 19, 436-43

- Gambrell, R.P., Collard, V.R., Reddy, C.R., and Patrick, W.H. (1977) - Technical Report D-77, Department of the Army, Washington, DC, 123

- Ganther, H.E., Goudie, C., Sunde, M.L., Kopecky, M.J., and Wagner, P. (1972) - Science, 175, 1122-4

- Gibbs, R.H., Jorosewich, E., and Windom, H.L. (1974) - Science, 184, 475-7

- Gilbertson, M. (1974a) - Bulletin of Environmental Contamination and Toxicology, 12, 726-32

- Gilbertson, M. (1974b) - Canadian Field Naturalist, 88, 273-80
- Gillispie, D.C. (1972) - Journal of the Fisheries Research Board of Canada, 29, 1035-41
- Harada, Y. (1968) - in Kutsuna, M. (editor), Minamata Disease, Shuhan, Kumamoto, pp73-117
- Heisinger, J.F., Hansen, C.D., and Kim, J.H. (1979) - Archives of Environmental Contamination and Toxicology, 8, 279-83
- Hughes, H.L. (1957) - Annals of the New York Academy of Sciences, 65, 454-60
- Irukayama, K. (1977) - in Tsubaki, T. and Irukayama, K. (editors), Minamata Disease, Kondasha, Tokyo, pp2-56
- Jernelov, A. (1970) - Limnology and Oceanography, 15, 958-60
- Kahrs, F.R. (1968) - Cornell Veterinarian, 58, 67
- Kitamura, S. (1968) - in Kutsuna, M. (editor), Minamata Disease, Shuhan, Kumamoto, pp257-66
- Klaverkamp, J.F., Turner, M.A., Harrison, S.E., and Hesslein, R.H. (1983) - Science of the Total Environment, 28, 119-28
- Koeman, J.H., Van den Van, W.S.M., de Goeji, J.J.M., Tjioe, P.S., and Van Haaften, J.L. (1975) - Science of the Total Environment, 3, 279-87
- Kramer, H.J. and Neidhart, B. (1975) - Bulletin of Environmental Contamination and Toxicology, 14, 699
- Kudo, A. and Mortimer, D.C. (1979) - Environmental Pollution, 19, 239-45
- Kurland, L. and Firo, L.N. (1960) - World Neurology, 1(5), 320
- Leonzio, C., Focardi, S., and Bacci, E. (1982) - Science of the Total Environment, 24, 249-54
- Luoma, S.N. (1977) - Marine Biology, 41, 269-73
- Luoma, S.N. (1983) - Science of the Total Environment, 28, 1-22
- Mackay, N.J., Kazacos, M.N., Williams, R.J., and Leedow, M.I. (1975) - Marine Pollution Bulletin, 6, 57-61
- Martoja, R. and Martoja, M. (1978) - L'Antagonism Selenium-Mercure, Actualités de Biochimie Marine, Colloque GABIM, La Rochelle, 149-84
- Millard, G.E. and Herbert, I. (1981) - Environmental Pollution (Series B), 2, 265-74

- Miller, G.E., Grant, P.M., Kishore, R., Steinkruger, F.J., Rowland, F.S., and Guinn, V.P. (1972) - Science, 175, 1121-2
- Pentreath, R.J. (1973) - Journal of Experimental Marine Biology and Ecology, 12, 1-15
- Pentreath, R.J. (1976a) - Journal of Experimental Marine Biology and Ecology, 24, 121
- Pentreath, R.J. (1976b) - Journal of Experimental Marine Biology and Ecology, 24, 103-19
- Pentreath, R.J. (1976c) - Journal of Experimental Marine Biology and Ecology, 25, 51
- Perkins, J.W. (1971) - Geology Explained in South and East Devon, David & Charles, Newton Abbot, p196
- Phillips, G.R. and Buhler, D.R. (1978) - Transactions of the Fisheries Society, 107, 853-61
- Phillips, G.R. and Gregory, R.W. (1979) - Journal of the Fisheries Research Board of Canada, 36, 1516-9
- Pillay, K.K.S., Thomas, C.C., Bondel, J.A., and Hyche, C.M. (1972) - Environmental Research, 5(2), 172-81
- Shultz, C.D., Grear, D., Pearson, J.E., Rivers, J.B., and Hylin, J.W. (1976) - Bulletin of Environmental Contamination and Toxicology, 14, 230-4
- Sims, R.R. and Presley, B.J. (1976) - Bulletin of Environmental Contamination and Toxicology, 14, 320-4
- Takeuchi, T. (1970) - in International Conference on Environmental Mercury Contamination, Ann Arbor, Michigan, 30th September - 2nd October, 1970
- Takizawa, Y. (1977) - in Tsubaki, T. and Irukayama, K. (editors), Minamata Disease, Kondasha, Tokyo, pp78-95
- Thornton, I., Watling, H., and Darracott, A. (1975) - Science of the Total Environment, 4, 325
- Tsubaki, T. (1977) - in Tsubaki, T. and Irukayama, K. (editors), Minamata Disease, Kondasha, Tokyo, pp57-78

CHAPTER X
SAFETY MEASURES AND POLICIES

A) FORMULATION OF POLICIES AND STANDARDS

The problem of mercury poisoning is not new but it is only recently that society has become aware of the problem. Long before the present concern about mercury poisoning, there were many documented cases of mercurialism, both acute and chronic, resulting from a variety of industrial and domestic exposures. The problem is accentuated by the fact that there are diverse and widespread man-made sources which release mercury into the environment. In the USA alone, the total discharge to the atmosphere will increase from the 1975 value of 473 metric tons to 615 metric tons in 2025 (figure 10.1). There are about 18 metric tons of mercury in municipal sewage sludge and this will increase to 37 metric tons by 2025. The chemical and physical forms of mercury are such that it moves readily from one environmental medium to another and is readily accumulated in sediments, soil, and living organisms. Furthermore, even if all mercury discharge to the environment was suddenly stopped, the residual already produced by fast activities, if not removed or permanently stabilised, will present a mercury problem for years to come.

The joint, FAO/WHO, committee on Food Additives (1972) and the OECD Sector Group on the Unintended Occurrence of Chemicals in the Environment (1974) limited the upper value of daily intake of mercury (as methylmercury) to 0.3mg for an adult. This corresponds to 0.3 μg Hg/g of fish.

In Japan, the Ministry of Health and Welfare set the weekly tolerable intake limit at 0.25mg of mercury (0.17mg of methylmercury) for a man of 50kg in weight. The national limits for mercury in fish and shellfish in some countries are as given in table 10.1.

MERCURY CONTROL IN RIVERS AND ESTUARIES IN EUROPE

The control of mercury in marine environments around Europe comes under the responsibily of the EEC Dangerous Substances Directive (CEC, 1976; Schubert, 1977). An environmental quality standard (EQS) of 0.3mg of Hg per kilogram of wet fish flesh has been set. The primary standard is derived from the minimum body burden of methylmercury in man which has been found to be associated with clinical signs of mercury poisoning. The minimum symptom figure, specified by UNEP/WHO (1976), corresponds to an intake of 200 to 500μg methylmercury per day for a 70kg man. An FAO/WHO expert group (1974) has recommended a provisional tolerable weekly intake of 300μg (i.e. 0.3mg/kg) of mercury, of which not more than 200μg should be in the form of methylmercury.

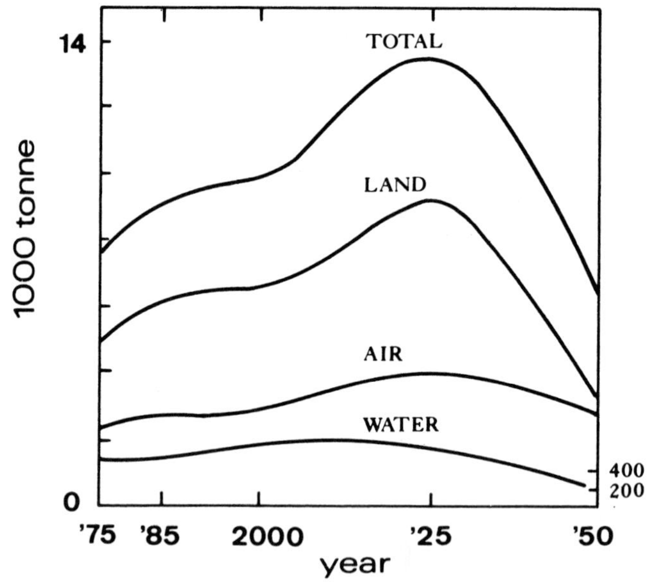

Figure 10.1 Estimated Weight of Mercury in Air, Land, and Water

However, symptoms of chronic mercury poisoning are very unlikely to be observed until blood levels reach the upper end of the WHO range for symptoms, i.e. 50μg/100ml (Krenkel, 1979).

Fish in the diet is regarded as being the most common source of mercury. The dietary studies involve the identification of a 'critical group' of consumers for quantification of the relevant parameters. Haxton et al (1979) and Hunt and Shepherd (1980) have undertaken duplicate dietary studies in order to identify the critical groups.

Table 10.1 National Limits for Mercury in Fish in Some Advanced Countries

Year	Country	Limit (ppm)
1967	Sweden	1.0
1968	Canada	0.5
1968	USA	0.5
1973	Japan	0.4

The formulation of policies and standards in two mercury-conscious countries (i.e. Sweden and the USA) are discussed below as models for a general approach to the problem.

POLICIES IN THE USA vis-à-vis SWEDEN

a) AQUATIC LIFE

Human beings accumulate mercury mainly by eating fish. The Food and Drug Administration, US Department of Health Education and Welfare has set "an 0.5ppm mercury interim guideline for fish" (Kolbey, 1970). If fish contain approximately 0.5ppm of mercury, then the daily intake limit of 0.03mg mercury would be reached by eating 60g of fish per day.

The average consumption of fish in Sweden is about 30g of fish per day (Berglund, 1971). Sweden has adopted a limit of 1ppm of mercury in fish. Berglund et al (1970) have observed that, with a level in fish corresponding to 0.5ppm, some people with extremely high fish-consumption might reach the lowest toxic level, and about one tenth of the population might reach the highest acceptable level.

The Swedish Board of the National Institute of Public Health, in consultation with the Swedish National Board of Health and Welfare and the Swedish National Veterinary Board, concluded that an acceptable level of mercury (as methylmercury) in human blood is 0.02μg/g, corresponding to a daily intake of mercury (as methylmercury) of

0.03mg.

The people of the US eat an average of two or three meals of fish per week. Assuming that the average two or three meals contain 0.5ppm of mercury, then the total intake of mercury per week is 0.22mg. Therefore, assuming no other intake, the 0.5ppm guideline of the Food and Drug Administration of the USA, is quite reasonable for the maintenance of public health. Canada also accepted this limit. The Study Group on Mercury Hazards (1970) concluded that, "on the basis of their examination of their experience in Sweden and Finland, the interim Food and Drug Administration (FDA), USA, guideline of 0.5ppm in fish is, for the present, a sound basis for the protection of public health. However, the margin of safety may not be large, and residual uncertainties remain in regard to possible hazards where large amounts of contaminated fish are eaten". The Food and Drug Administration of the USA has banned the entry of fish, containing mercury residues greater than 0.5ppm, into the market.

The Federal Water Pollution Control Act (1972) regulates mercury discharge to water. Because of the restriction imposed by this act, and a watch kept by the US Environmental Protection Agency (EPA), mercury discharges to water are likely to come down from the 60 tonnes of 1975 to 50 tonnes in 2025.

b) DRINKING WATER

The Bureau of Water Hygiene of the United States Public Health Service proposed a standard of 0.005ppm for mercury in drinking water (Bureau of Water Hygiene, 1970).

In 1970, it was reported that the Great Lakes area of the USA were contaminated with mercury, but it soon became evident that the pollution was not limited to the Great Lakes but was of national scope.

Therefore, the Regional Directors of the Federal Water Quality Administration, USA, systematically checked all known users of mercury in order to ascertain whether they discharged mercury into the aquatic environment. Natural waters were checked for mercury contamination and priority was given to the determination of (1) the extent and intensity of the contamination, (2) whether unacceptable residues in fish might be present, and (3) whether water supplies might be dangerous.

The Safe Drinking Water Act (as amended, 42 USC, 300f et seq) was passed in 1974.

Comprehensive National Interim Primary Drinking Water Regulations (NIPDWR) (40 CFR, 141, 1975) were published in December, 1975. The regulations fixed the limit

of mercury at 0.002mg/l.
In 1975, interim regulations for bacteria and turbidity, ten inorganic and six organic chemicals and radio-nucleides were promulgated by EPA for the protection of drinking water. In 1979, the National Secondary Regulations for substances which affect the aesthetic quality of water was introduced. To that year's list, trihalomethane was added. The United States are engaged in comprehensive revisions of the National Primary Drinking Water Regulations (Cotruvo, 1981).

c) INDUSTRIAL DISCHARGES

The Resources Conservation and Recovery Act, passed by the US Congress in 1976, established quantitative requirements for a permit system for the disposal of hazardous materials from industrial sources, but not those from household individuals, in landfills. However, it, "does not interpret the intent of Congress to include the regulation of individual home-owners, due to the small quantities of hazardous waters which they may generate" (Federal Register).
However, the Resources and Recovery Act does, in another section, provide for the promulgation of guidelines for the management of municipal landfills by state and local authorities.
This act stipulates, "the maintenance of operation of such facilities and requires such additional qualifications as to ownership, continuity of operation, training of personnel, and financial responsibility as may be necessary or desirable" (US Code Service, December, 1976, p4973).
It is known that almost half of all of the mercury discharged into the environment is located in municipal landfills. One approach to reducing this would be to levy redeemable deposit charges on primary batteries and other consumer products containing mercury. In the case of primary batteries, recycling costs are relatively low.
The Federal Register, 1975 (p48302) indicates a mercury discharge limit of 2300g/day from any industrial plant, mainly mercury smelters, chlorine producers, and sludge incinerators, etc.
In the USA, the discharges to land are expected to increase from 1092 tonnes in 1975, to 1336 tonnes in 2025 (figure 10.1).

d) DISCHARGES TO THE ATMOSPHERE

The US Environmental Protection Agency (EPA), in an effort to reduce emissions of mercury to the air, has adopted the following actions: (1) conduction of a more thorough and detailed identification of emission sources, (2) development of accurate emission factors for each category of source, and (3) adoption or development of control systems which are able to reduce emissions to below the level which would be hazardous.

Mercury emissions to the atmosphere, and their effects, are not well known. It has been estimated that, in the USA alone, about 1000 tons of mercury are released into the atmosphere and, without any proper controls, this will increase by 5% per year.

The regulations for mercury in air are few. The greatest air emission sources in the USA are volatilised mercury from water-based paints, fuel-burning, and volatilised mercury from incinerated solid wastes.

Using the US Clean Air Act (1970) and its amendments (CAAA) of 1977, the Federal Administration can levy effluent charges, input charges, and/or financial penalties for failing to meet either performance standards or equipment installation requirements. The underlying idea was to make it cheaper for industry to comply with the regulations than to ignore them or opt for the cheapest-possible technical fix. The Clean Air Act (1970) also incorporated a "technology-forcing" element into air pollution control with stringent standards and actions to ensure that appropriate equipment is installed.

The CAAA, "involved considerable expenditure on manpower and financial resources by both government and the private sector. For example, in 1977 $22.3 billion were spent by both the public and private sectors in attempting to meet air pollution standards" (Brady & Cunningham, 1981).

e) PESTICIDE CONTROL

Mercury pesticides, which are widely used in agriculture, can be a significant source of mercury to the environment. In Sweden, when certain bird populations decreased drastically, investigation revealed the presence of large amounts of mercury in the dead birds. Furthermore, increased amounts of mercury were observed in fish as well as in polluted waters. The death of birds was related to the extensive use of methylmercury dicyanodiamide as a fungicide in Swedish agriculture. The use of methylmercury compounds as fungicides has been banned in Sweden and, since then, the mercury content of birds and eggs has decreased substantially.

In order to bring about pesticide control, the United States issued four notices of cancellation (during 1969-71):

1 1st February, 1969 - Cancellation of phenylmercuric monoethanol ammonium acetate compounds for use on apples and strawberries, phenylmercury nitrate compounds for use on almonds and prunes, and phenylmercury urea compounds for use on barley, corn, cottonseed, flax, oats, peas, rice, rye, sorghum, and wheat.

2 26th February, 1970 - Cancellation of phenylmercuric acetate or phenylmercuric ammonium acetate compounds for use on millet, rye, and sugarcane

3 12th March, 1971 - Cancellation of hydroxymercuric chlorophenol compounds for use on snap beans, sweet corn, cowpeas, flax, peanuts, peas, potatoes, seedlings, soybeans, sweet potatoes, and velvet beans, hydroxy-mercurinitrophenol hydroxymercurichlorophenol compounds for use on potatoes and sweet potatoes, methylmercury 8-quinolinolate compounds for use on apples, and phenylmercuric acetate or phenylmercuric ammonium acetate compounds for use on apples, cherries, peaches, strawberries, and sugarcane.

4 8th August, 1970 - All mercury products bearing claims and/or directions for use as slimicides or algicides, or for use in laundering were cancelled. This action was taken because the use of these mercury compounds results in water contamination which is potentially injurious to man and his environment.

On the 9th March, 1970, all pesticide products containing alkylmercury for use as seed treatments were suspended while, on the 7th October, 1971, algicides for use in swimming and paddling pools and for industrial use were suspended.
On the 18th February, 1976, the US Environmental Protection Agency ordered an immediate ban on the production of virtually all pesticides containing mercury. This preventative measure banned the manufacture of all mercurial pesticides used as bactericides and fungicides in paints, varnishes and lacquers, and prohibited the use of such pesticides on turf, including golf courses, and in the treatment of seeds. The EPA estimated that the ban would reduce the mercury presently entering the environment by about 98.5%

DECISION MAKING IN SWEDEN

Lundqvist (1974) reports that mercury pollution became the first large biocide problem in Sweden, causing legal battles and policy-framing.
During the 1960's, there were two destructive cases of mercury contamination; one concerning birds and eggs, and the other concerning fish. The bird problem did not affect the average consumer, except for the ecological and conservancy aspects.
The National Poisons and Pesticides Board of Sweden had to go through a legal battle on whether the agricultural practice of treating seed grain with alkylmercury

compounds should or should not be allowed to continue.

The seed grain treatment producers, the farmers, the pulp and paper and the chlor alkali industries, became the subject of controversy. Certain waters were blacklisted by analytical organisations. However, administrative decisions were severely hampered by the multitude of agencies having some kind of responsibility for the regulation of public health and of substances in food products.

During the most intensive stages of the mercury-in-fish issue, the press, radio, and television were extensively used by the scientists; both to publicise research reports and to attack each others statements.

MERCURY IN BIRDS

From 1957 to 1962, Dr Karl Borg and his group collated the reports of investigations, carried out at the National Veterinary Institute, on wildlife found dead in the environment. In about 1962, the scientists of the National Resources Committee of the Swedish National Science Research Council joined in such investigations, vis-a-vis mercury poisoning. In 1963, Dr Gunner Otterlind at the Sea Fisheries Laboratory, whose research was sponsored by the Ornithological Association, took a very active part in this study.

The investigations were further enhanced by scientists such as Dr Stig Tejning at the Institute of Occupational Medicine in Lund, Dr Torbjorn Westermark at the Royal Institute of Technology, and Alf Johnels at the National Museum of Natural History. All of these scientists revealed a correlation between the level of mercury in birds, and the treatment of seed grain with alkylmercury compounds.

In 1965, the National Institute of Public Health (NIPH) and the Institute for Seed Control were experiencing some difficulty in identifying the criteria for recommending mercury levels to the deciding agencies, i.e., the National Veterinary Board and the National Board of Medicine.

Other views were contributed, to the controversy over control, by the Swedish Ornithological Association, the Swedish Hunting Association, the Swedish Society for the Protection of Animals, and the Swedish Association of Seed Grain Producers.

MERCURY IN FISH

The issue of mercury in fish is more a question of public health than one of ecology.

In 1964, scientists from the Natural Resources Committee, the National Museum of

Natural History, and the Royal Institute of Technology, began to study fish samples from several Swedish fresh water bodies. As a result, an environmental threat from mercury pollution was brought to light.
On matters of toxicity and effect on human beings, several departments at the University of Stockholm made important contributions to scientific knowledge.
The investigations indicated that three sources supplied mercury to Swedish waters. The paper-mills used phenylmercury in order to prohibit the formation of mucus in the paper machines. The pulp industry used phenylmercury to protect wet mechanical wood pulp from the attacks of mould fungi. The National Poisons and Pesticides Board was to regulate the activities in these two cases.
When the Environmental Protection Act (EPA) became law in July, 1969, the chlor-alkali producers; the third source of mercury, came under strict regulation.
When the National Institute of Public Health found dangerous concentrations of hazardous substances in food products, it was to inform the authorities responsible for administrative decisions concerning food substances and matters of public health. The Institute was also to give policy recommendations to these agencies, i.e. to the National Veterinary Board (for substances in food) and to the National Board of Medicine (for matters of public health). The final decision on maximum allowable concentrations of mercury in food thus had to be taken jointly by these two agencies. Since the issue of mercury in fish touches directly upon the Swedish fisheries, the National Fisheries Board is also involved.

B) SAFETY MEASURES AND HYGIENE

Several countries have adopted mercury safety standards. An International Committee (Stockholm, 1969) suggested a level of 0.05mg Hg vapour/m^3 of air to be the maximum concentration allowable for industrial exposure. The committee also reported that non-specific symptoms such as insomnia, shyness, nervousness, and dizziness increased in frequency if the exposure was higher, although exceptions were observed on an individual basis.
The US Public Health Service has set a tentative upper limit of 5ppb of Hg in drinking water. A similar limit is also specified in the USSR.
The Food and Drug Administration (USA) discards all pods with more than 0.5ppm of methylmercury as being undesirable for a single meal. The approved concentration of methylmercury in the USA and USSR is 5ppm of methylmercury for a single meal. The

approved concentration of methylmercury in the USA and USSR is 5ppm of methylmercury for a single meal. The approved concentration of methylmercury in the USA and USSR is 5ppb. Japan has set the limit on the amount of methylmercury in industrial discharges at 10ppb, while the USA has set the higher limit of 50ppb. India has yet to adopt similar safety standards for mercury use and discharge. In India, the chlor-alkali industries alone discharge 1300 tonnes of mercury into the environment each year and therefore it seems to be high time that curbs are imposed which are similar to those in other countries.

In the USA, a permissible limit of 0.5ppm of mercury had been established for foodstuffs. Therefore when, in 1970, Hg analyses of fish from Lake Ontario (Canada and USA) revealed quantities which were very much in excess of this, e.g. about 7ppm, health authorities became alarmed. Subsequent analyses of canned tuna fish also showed some cases of Hg contents which were in excess of the permissible limit, with the result that great quantities of canned tuna fish were taken off the shelves. Swordfish were found to accumulate high concentrations of mercury and its sale has therefore essentially ceased in the markets of North America. The north-eastern USA have upper limits of mercury of the order of 0.04 ± 0.02ppm in surface soils, about 0.06ppb in water, and a few nanograms per cubic metre in the atmosphere (CEK report, 1970). This report also shows that methylmercury enters organisms at the beginning of the food chain via the ingestion of fine sediments, and that almost all (98%) of the mercury found in animal tissues is in the form of methylmercury.

In order to keep the polluting industries running, and simultaneously reduce mercury contamination of the soil, water, and air, the uses of mercury compounds are to be regulated and the effluents treated. In the USA, the respective industries reduced the discharge of Hg by some 86% within a few months of alarming Hg concentrations being found in the fish and sediments of Lake Ontario and other areas in 1970.

The low rate of excretion of mercury by fish, and the continuing supply of mercury from bottom sediments to the water give little hope for the lowering of mercury levels in fish. The work carried out by the Swedes confirm this. In Sweden, even after the imposition of a ban on mercury use as long ago as 1966, the mercury concentration in the pike of most lakes has remained constant. Only in three lakes did the mercury level in pike decrease significantly. Rivers have a better chance due to the continual flushing action. The mercury levels of salmon placed in cages below former sources of mercury in some Swedish rivers showed a considerable improvement within three years

(Study Group on Mercury Hazards, 1970).

Jernelov (1969) has calculated that about 10 to 100 years would be required to remove the mercury from the bottom of the lakes by the process of methylation. In Minamata Bay, after the cause of the pollution had been identified and estimated, the mercury levels in shellfish dropped from 85ppm to 10ppm over a two-year period and remained constant for five years (Irukayama, 1966).

The following approaches to the decontamination of mercury-contaminated waterways have been suggested by Swedish workers:

1 Introduction of oxygen-consuming materials in order to create permanent anaerobic conditions in the sediments, thereby reducing methylation.

2 Increase of the pH of the sediments in order to favour demethylation and increased volatilisation.

3 Covering of the sediments with fresh, finely-divided materials of high adsorptive ability (e.g. quartz and silicates).

4 Covering of the sediments with inorganic inert materials of any type.

5 Removal of mercury-bearing sediments by dredging or pumping (Study Group on Mercury Hazards, 1970).

Experiments have been conducted in Sweden in order to evaluate the covering of sediments by layers of inorganic layers of varying thickness (0 to 20cm), with and without ***Tubificidae*** (oligochaete worms) and ***Anodonta*** (a bivalve). These studies have shown that: (1) in the absence of the former, methylmercury in fish accumulates only when the sediments are uncovered, (2) in the presence of large populations of these worms, fish accumulated methylmercury when the covering layer was less than 2cm thick, and (3) in the presence of ***Anodonta***, which stir up the sediments, leakage of methylmercury occurred if the covering layer was less than 9cm thick.

The effect of ground silicates upon the uptake of mercury by fish from sediments contaminated with metallic mercury, ionic mercury, and phenylmercury, was evaluated by Swedish workers who found that there was no reduction in uptake when the pollutant was phenylmercury. However, a decrease in uptake amounted to a factor of two when inorganic mercury was the pollutant (Study Group on Mercury Hazards, 1970).

The Swedes, on two occasions, dredged the sediments from affected waterways and subsequently found that the mercury concentration in fish became higher than before. The reason proposed was that, when the anaerobic sediments in which the mercury was bound up as the sulphide were released to the air, the mercury was converted to the ionic form and the sulphide was converted to sulphate. The bacteria were then able to convert the mercury to the toxic methylmercury form, and rain subsequently washed the methylmercury out into the waterway, where it was absorbed by the fish.

The method of removing mercury from industrial outfalls is based upon the precipitation of the soluble mercury salts as insoluble mercuric sulphide by the use of sodium sulphide (Spence, 1971). The pH of the waste liquids containing mercury salts is adjusted until the waste solution is basic since, under acidic conditions, the addition of sodium sulphide will liberate toxic H_2S gas. Next, sodium sulphide solution is added and the liquid is allowed to stand for several hours. The black precipitate of HgS settles, and the filtrate can be poured into the streams.

For good mercury hygiene in dental practice, the following recommendations have been proposed by Rupp and Paffenberger (1971).

a Mercury should be stored in unbreakable, tightly-sealed containers.

b Any inadvertent spillages should be confined to an easily cleaned tray.

c The surgery should have seamless flooring which extends two inches up each wall.

d The cavity surfaces should be coated with varnish.

e All amalgam scrap should be salvaged and kept in a tightly-covered container.

f Work should be carried out in well-ventilated spaces.

g The indiscriminate use of mercury-containing solutions should be eliminated.

h The heating of mercury or amalgam should be avoided.

i Water sprays and suction should be used when grinding dental amalgams.

j Ultrasonic amalgam condensers should not be used.

The potential of methylmercury as a mass killer has become well-known following the Minamata and Niigata episodes, and alarm has been accentuated by the discovery that methylmercury can be synthesized from other mercury compounds by microbial and other agencies, and that, once formed, it is highly resistant to biodegradation. With the increase in awareness of this aspect of the toxicity of mercury and its compounds, checks are being imposed on the use of mercury and the disposal of mercury-bearing wastes, etc, from industrial and agricultural organisations. In Sweden, such controls have already produced beneficial effects and it is expected that other countries will adopt similar measures.

C) LEGISLATION ON TOXIC ELEMENTS

JAPAN

Legislation: the principal laws which govern the utilisation of mercury or the reduction of its emissions are:

- The Fundamental Law against Environmental Pollution (1967, 1970)
- The Law on the Prevention of Water Pollution (1970), which has the following norms for mercury discharges into water and effluents:
  (i) mercury in any form should be untraceable
  (ii) methyl- or ethyl- mercury should be untraceable

As regards the industrial discharge of mercury, the law dealing with the removal of discharges and the law on public health (1970) indicate the treatments which are to be applied:

i The mercury content of the discharges should be removed, as far as possible by heat treatment, or

ii It has to be transformed into insoluble sulphides which are either pressed into a concrete block and sunk in the ocean at prescribed locations (depending upon the law concerning sea pollution) or buried in an isolated region away from public water supplies. With regard to the properties of the concrete block, a series of tests have been conducted for determining the resistance to squeezing out of the mercury from the block. The application of this law is usually overseen by the Ministry of Public Health and the Environmental Protection Agency.

Other relevant laws (and the appropriate authorities) are:

- The Law on the Utilisation of Chemical Products in Agriculture (1948, 1970). Ministry of Agriculture, Water, and Forests, Environmental Protection Agency
- The Law on the Pollution Control of Cultivable Land (1970). Ministry of Agriculture, Water, and Forests, Environmental Protection Agency

- The Law on the Control of Toxic and Harmful Substances (1960, 1970). Ministry of Public Health

- The Law on the Hygiene of Food Products (1947). Ministry of Public Health

- The Law on Sewer Construction (1950, 1970). Ministry of Construction

- The Law on Pharmaceutical Products (1960). Ministry of Public Health

- The Law on Security in Mines (1949). Ministry of Commerce and Industry

- The Law on the Protection of Aquatic Resources (1951). Ministry of Agriculture, Waters, and Forests

- The Law on Sea Pollution Control (1970). Ministry of Transport

- The Special Law on Public Activities for Specified Health Risks due to Environmental Pollution (1969). Environmental Protection Agency

- The Law on Arbitration or Litigation arising from Environmental Pollution (1970). Arbitration Committee of the Prime Minister's Secretariat

- The Law on the Attribution of Allocations to Public Works Entrepreneurs for the Reduction of Pollution (1970). Environmental Protection Agency

Principal Measures taken against the Use of Mercury and Mercury Compounds with regard to Pollution in Japan:

1 Prevention (1970) of the utilisation of all mercury-based organic compounds, such as fungicides in paint manufacture, as a result of directives from the Ministry of Industry and International Trade.

2 Prescription (begun in 1966, finalised in 1968 - law on the utilisation of chemical products in agriculture) of the use of mercury-based pesticides in any agricultural operations except the treatment of seed, and the prevention (begun in 1968, finalised in 1970 - law on the use of chemical products for agriculture) of the admixture of any mercury pesticide in the manufacture of any product used other than for the treatment of seed.

3 Ending (1971) of the manufacture and sale of contraceptive products containing mercury, as a result of a directive based upon the law on pharmaceutical products.

4 Publication (1958) of a report on the control of industrial effluents (which is currently incorporated into the 1970 law on measures against water pollution), which fixed the admissible limits of the mercury level in the effluents. This publication has led to the closure, by a voluntary decision of the management, of a large number of factories which used mercury. The managers felt that their

economic and technical constraints would not permit the satisfaction of the limits fixed for the effluents.

5 Restrictions and warnings on the catching, sale, and consumption of fish (Directives from the Ministry of Public Health; Law on the Hygiene of Food Products)

SWEDEN

Legislation: the principal law controlling mercury pollution is:

- The Law on Environmental Protection (1959)

Other related laws, and the associated administrative bodies are:

- The Law on Pesticides (1962). National Office of Pesticides and Poisons (now known as the Products Control Service, Environmental Protection Agency).
- The Decree on Pesticides (1963). National Office of Pesticides and Poisons (now known as the Products Control Service, Environmental Protection Agency).
- The Law on Poisons (1963). National Office of Pesticides and Poisons (now known as the Products Control Service, Environmental Protection Agency).
- The Decree on Poisons (1963). National Office of Pesticides and Poisons (now known as the Products Control Service, Environmental Protection Agency).
- The Law on Water (1918-21). National Agency for the Protection of Water (now under the auspices of the Environmental Protection Agency).
- The Law on the Surveillance of Aquatic Zones (1956). Environmental Protection Agency.
- The Law on Water Works and Water Purification (1970). National Agency for the Protection of the Environment.
- The Law on Food Products (1971). National Food Agency.
- The Decree on Food Products (1971). National Food Agency.

Principal Measures taken against the Use of Mercury and Mercury Compounds with regard to Pollution in Sweden:

1 Certain restrictions (early 1960's - Law on Pesticides) upon the use of

mercury-based products for the treatment of seed.

2 Prohibition (decision taken in 1966 within the ambit of the Law on Pesticides) of the use of alkylmercury compounds for the agricultural treatment of seed.

3 Prohibition of the use of all organomercuric compounds in the paper and paste industry. In particular, prohibition of their use for the development of viscous products (1966), and for the impregnation of paper paste (1967 - Law on Pesticides).

4 Prohibition (begun in 1967 - Decree on Food Products) of the sale of fish from some 70 sea areas.

5 Obligatory reduction (1967 - Laws on Water and the Surveillance of Aquatic Zones) of emissions to water and air by industries using mercury; especially those producing chlorine and soda by means of mercury electrodes.

CANADA

Legislation: the laws governing mercury pollution are those on fisheries, pesticides, food products, and medication.

- Law on Fisheries (1952, amended 1970). With regard to fishing zones and the pollution of water, the law requires (article 3) that no person should dump, or permit to be dumped, in water inhabited by fish, any substance, or water containing substances, which are capable of affecting the quality of water from the point of view of the survival of fish or their consumption by man. Measures have been taken to test these regulations on the substances, their levels or admissible concentrations in the effluents dumped into the waters, and the treatment processes for changing the quality of liquid effluents. Within the ambit of this law, mercury present in the liquid effluents coming from chlorine and soda factories has been cited as a prohibited substance and, in 1972, the "regulations on the control of concentration of mercury in the various effluents from the manufacture of chlorine and caustic soda" have been incorporated as an amendment to article 33 of this law. This amendment prescribes 0.005lb (1.82g) of mercury per ton of chlorine produced to be the maximum daily dose of mercury in the liquid effluents, coming from the electrolytic installations treating alkali chlorides, which are dumped into waters inhabited by fish. Each and every factory or plant should monitor its mercury effluents and maintain an inventory of all of the mercury that has been bought, stocked, used, and discharged.
  The Federal Ministry of the Environment holds the responsibility for the application of the law on fisheries but, in certain cases, it delegates its powers to regional administrations. These regulations would be revised within two years.

Other laws against mercury pollution are:

- the Law on Pesticides (1968-9); ratification of mercury components.
- the Law on Food Products and Medication (RSC-1970); prescription of mercury dosages.
- the Law on the Inspection of Fishery Products (1949, as amended); covers fish and fishery products, which circulate in inter-regional and international commerce, from the point of view of health and public hygiene.

<u>Principal Measures taken against the Use of Mercury and Mercury Compounds with regard to Pollution in Canada</u>:

1 Confirmation of all of the mercuric products proposed for treatment of cereal grains (excluding stocks which have already been sent to retailers); a decision taken in 1970 within the ambit of the Law on Pesticides.

2 Sixteen fishing zones have been totally or partially closed to commercial fishing, with notification and destruction of fish and fishery products exceeding the maximum prescribed rates (1970 - Law on Fisheries).

3 Instigating legal proceedings against certain bodies which are discharging mercury into their industrial effluents.

4 Reduction in the amount of mercury dumped by electrolytic installations using alkaline chlorides.

USA

<u>Legislation</u>: the principal laws concerning the protection of the environment against the introduction of toxic substances, including mercury, are:

- the Federal Law (1972) against the Pollution of Water
- the Law on Industrial Refuse (article 13 of the 1899 Law on Waters and Ports)
- the Law on the Health of the Climate (1963)
- the Law on Food Products, Medications, and Cosmetic Products (1938)
- the Federal Law on the Protection of the Environment (1972), which replaced the federal law on insecticides, fungicides, and rodenticides of 1947 (Federal insecticides, fungicides, and rodenticides act - FIFRA)

<u>Principal Measures taken against the Use of Mercury and Mercury Compounds with regard to Pollution in the USA</u>:

1 The Food and Drug Administration (FDA) decided (1965) that the packing paper in cartons coming into contact with consumable food products should be free from mercury.

2 Withdrawal (1969, 1970, 1971, FIFRA) of a series of organic compounds based upon mercury and used as fungicides in agriculture.

3 Suspension (1970, FIFRA) of all mercury products based upon alkylmercury in their utilisation for the treatment of seed.

4 Withdrawal (1970, FIFRA) of all mercury-based products used as antislime agents (slimicides) or algicides, as agents for the destruction of musilageneous (viscous) products, and used in laundries. However, on the 7th November, 1971, the Environmental Protection Agency decided against the inclusion of this component.

5 Under the aegis of FIFRA, various procedures were adopted on the 22nd March, 1972:

a suspension of:
- all alkali mercury-based pesticides for any use
- all mercuric pesticides for the treatment of feed rice
- all mercuric compounds used in laundries
- all mercuric compounds used as preservative agents (antifouling agents)

b suspension of all other uses declared and involving mercury; all mercury compounds used as fungicides that have not yet been withdrawn or suspended; all mercury compounds which are used as preserving agents or moisteners in paints, adhesives, products based upon amidon, gums, emulsions, and other similar products; in compositions which are to be used on surfaces; on tissue, textiles, and fibres; for the treatment of furniture and masonry; in moisture-preventing agents in papers, cartons, and sintered fibre boards; in tanning; in the specimen preparation of milk and urine; in the fight against hay-fever; in recipients for food products, animal furs, and other products used by man; in food product cultures; tobacco, and fur products, on trees, bushes, and flowers; as water baths for bulbous plants and cereal grains, and for the treatment of sorghum or coco-fibres; cellulose sponges, and other similar products.

6 Judicial proceedings by the Federal Administration against industries identified as mercury pollutants (Federal Water Pollution Control Act and Refuse Act). A large number of such polluters have already been brought to book or are awaiting administrative decisions.

7 Confiscation (1970 onwards; Law on Food Products, Medication, and Cosmetic Products) of polluted fish and crustaceans.

8 Replacement (1970 onwards; Federal Water Pollution Control Act) of mercury joints in sewage water treatment plants.

9 Prohibition (1970 onwards; following the national laws) of Fishing and Destruction of Fishery Products by various state authorities.

10 Administrative procedures for fixing atmospheric emission norms (1971 Clean Air Act)

FRANCE

Legislation: reproduced below is the text of the circular, dated 13th May, 1974, concerning mercury emissions from electrolytic alkaline chloride plants.

- Measures taken for the reduction of mercury pollution. This text is an extract of a circular issued by the Ministry of Health, and dated 26th March, 1975.
- Measures taken in the context of the production of chlorine using mercury electrodes. This type of production makes up some 30 to 40% of the pollution caused by mercury in France, which amounts to 50 to 1000 tonnes of mercury per year.

Circular dated 13th May, 1974, regarding mercury pollution by plants producing alkaline chlorides via the electrolytic process.

Article 1 -

the installation of electrolytic plants producing alkaline chlorides using the mercury process would be authorised only if proper measures were taken, right from the planning stage, to ensure that the mercury levels inherent in that process are less than the values defined in article 2, or if these cases are compatible with those of other existing plants or with the regulations defined at that time.

Article 2 -

a the losses to any waste water should be lower than 2g/ton of chlorine produced; the average value being calculated on the basis of the annual production and rejection levels.

b the losses to the atmosphere should be less than 7g/ton in ducted and non-ducted chlorine.

c the losses to finished products should be less than 2g/ton of chlorine produced, and made up thus:
- in a 47% caustic lye (leach), 1.2g/ton of chlorine, amounting to 0.5mg of mercury per kg of lye
- in hydrogen, 0.8g/ton of chlorine; amounting to 2.5mg of Hg per cubic metre of normal hydrogen.

These values are calculated on the basis of monthly production figures.

Article 3 -

- every precaution has to be taken to ensure that the mercury present in solid or semi-liquid refuse does not contaminate the natural surroundings.
  In particular, this refuse should be stored in air-tight containers and elimination or recuperation should not take place until after exhaustive treatment for mercury recovery by the plant authorities concerned or by enterprises which carry out such recuperation under the guidance of the Ministry of the Environment and National Protection.

Article 4 -

precautions have to be taken at the time of conception and installation of the equipment, as well as during actual production, in order to limit non-measurable losses of mercury.
The foundation of the plant should be impervious to water, and the sewage network should collect all of the effluents on the surface, including accidental spillages.
In practice, the floors, basements, and sewage canals should be frequently washed, and the water used for this purpose be collected for treatment as waste. All waste, either scrap or any other matter suspected to contain mercury, should be recovered. The handling of mercury should be reduced to a minimum and strict surveillance be maintained so as to prevent losses and thefts.

Article 5 -

the method of sampling, as well as the method of determination, of the output is to be decided in liaison with the factory inspector and the plant authorities.

The frequency of analysis is to be as follows:

- In waste water: daily analysis of a sample representing the average effluent.
- In soda: daily analysis of a representative sample of the production.
- In hydrogen and ducted gaseous effluents: at least one analysis per month
- In non-ducted gaseous effluents: one analysis per month.

These results and monthly production figures are to be kept in a register and produced when demanded by the factory inspector.
Conversely, the mercury position of the plant is to be reviewed every 18 months and communicated to the factory inspector.

COMMENTARY

Article 1 -

The risks presented by the accumulation of mercury in the food chain up to man, based upon the percentage in the natural surroundings - even if very low, would permit public authorities to prescribe a drastic reduction in the quantities of

mercury rejected into the surroundings and finally into marine water. It is in this context that the interministerial action committee on the natural environment demanded, on the 6th December, 1972, a reduction of 50% within 2 years, and a reduction of 90% within 5 years, of the amounts of mercury released into the seas.
It is particularly necessary to regulate strictly the mercury losses from electrolytic plants using alkaline chloride (and producing chlorine, soda, and sometimes potassium) and the cathodic mercury process.
In addition, certain zones appear to be already contaminated by mercury, whether due to natural causes or human activities.
The reduction of the emissions in these areas should be achieved at a much higher rate, and any erection of plants based upon the above-mentioned process should be refused permission. This should also hold true for those areas which are uncontaminated or less-contaminated if the need is felt to halt the above-mentioned contamination. Attention is drawn to conchytic zones and potable water sources.

Article 2 -

1 Waste water arises from a number of sources:

The process and washing water coming out of the electrolytic chambers,
Losses of pickling brine, including washing of the filters,
Condensation of the water used for washing hydrogen,
Condensation of waste water from the treatment of mercuric mud.

Internal measures, such as reviewing the layout of pickling brine circuits with the aim of reducing collecting circuits and improving their accessibility to washing and the collection of accidental flows, have to be taken into account for each of the circuits before recuperating the waste water for treatment.
After the recuperation, external treatment of the waste water has to be undertaken if the mercury content in this water exceeds the prescribed limits.
Cooling water and, in general, any water not containing mercury (except by accident) has to be collected separately, before final recuperation of the product.

2 Losses to the atmosphere from ducted sources such as the purifying gases from the entry and exit cells and the treatment vats for the mercury refuse and non-ducted losses, especially those that occur in the electrolytic chambers and which are drawn away by natural ventilation or forced out of the chambers.
It is proposed to duct these losses to the maximum possible extent at the end of treatment.
Precautions are to be taken against non-ductable losses by designing proper cells and hydrogen circuits, equipped with preventative and recuperative devices for excluding hydrogen gas.

3 If the finished products are transformed at the place of production itself, by the plant authorities, the mercury losses could be deduced from the derived products, provided that the excess mercury is effectively conserved and not dispersed during the transformations.
If the plant produces potash instead of soda, the mercury losses in the lye should be less than 0.5mg/kg of lye.

The losses of chlorine produced have not been taken into consideration as the percentage of losses in this case are negligible (about 0.01g per ton of chlorine).
The limiting values given in Article 2 should be progressively reduced within a period of 5 years as a result of the technical refinements achieved.
For hydrogen, in particular, the processes which are under development should, in the near future, permit the achievement of a level of less than 100ng per cubic metre. With regard to commercial hydrogen in the form of compressed gas, this value has already been obtained.

Article 3 -

Among the solid and semi-liquid wastes which are likely to contain mercury, those worth citing are mud coming from the purification of brine, mud coming from the treatment of water containing mercury, used (discoloured) activated charcoal, dirty scrap, and washing wastes.
The muds from various sources are to be stored only after treating them using processes which fix the mercury in the sources in the form of insoluble compounds.

Article 4 -

In chlorine units which use mercury there are, in addition to the measurable losses given in Article 2 and the losses which are found in certain categories of solid refuse, other non-measurable losses which are characterised by the fact that they can be qualitatively but not quantitatively measured.
The principal non-measurable losses are those due to infiltration into soil, losses due to scrap in contact with mercury, washing wastes, losses in handling, and theft. The prescribed measures described in Article 4 should permit them to be considerably reduced. Special training of personnel would be of help in this direction. The issuing of circulars on preventative measures to the personnel would also be of help in these cases.

Article 5 -

The values given in Article 2 are only average figures. The analytical method to be used for these checks is an atomic absorption method which is specific to the chlorine industry and which is known as the BITCL method. This has been the subject of inter-laboratory verifications throughout Europe, and has been published in the journal, 'Analytica Chimica Acta'. In the case of non-ducted gaseous effluents, this method could be replaced by a simplified one such as the CuI calorimeter, and the control of mercury losses could, in principle, be based upon calculations of the level of mercury in the plant atmosphere, measured by sampling the yield resulting from renewing this atmosphere. The method used for determining the amount of refuse which could occur (as well as its nature, volume, and mode of emission) must be spelt out before commissioning the plant.
The verification of the results recorded in the register should reflect the correct methodology of the measures, and the representiveness of the results themselves.
The factory inspectors may ask for the verification of these results by a mutually agreed laboratory, and the expenses of this are to be borne by the plant authorities.
The establishment of a mercury balance sheet should be based upon differences

in the inventories, established by using radioactive tracers.

Two circulars issued in May, 1974 are applicable to this source of pollution:

1 The first one is aimed at the new plant which should, right from the construction stage, use those processes which will permit a 90% reduction in the normal emission of plants, as compared to those using earlier processes.
The circular also prohibits the installation of plants using this process in particularly sensitive zones. It was for this reason that the latest FOS plant would be using diaphragm cells devoid of mercury.
The second circular establishes a programme, for the reduction of pollution caused by earlier plants, which is comparable to the CIANE programme of the 6th December, 1972.

2 Measures taken with regard to other sources of mercury pollution. The other sources are much rarer and are the subject of specific programmes adapted to each case.

The more important measures already taken are:

- The withdrawal of all homologous agricultural fungicides containing mercury.
- Inclusion in the 'C' list of all poisonous substances in paints containing mercury salts (marine paints), according to Article R5168 of the Code of Public Health.
- Actions to be taken in adopting procedures for the collection and recuperation of objects containing mercury (batteries).
- Prohibition of the use of fungicides and biocides based upon mercury in the paper and paste industry.

The measures taken at national level have been enforced at certain Mediterranean locations. Therefore, the permanent secretariat for industrial pollution problems in the FOS region has established very strict dumping rules for industries in that region.
Thus, mercury dumping in the region is limited to 0.1mg/l.

The decisions of CIANE are usually adhered to, and the carefully planned programme is being followed. Such a programme is the first to be established in Europe.
Taking into account the cumulative nature of mercury, it is all the more necessary that those national measures be accompanied by international action.

INDIA

REGULATION AND CONTROL OF TOXIC METALS

The legislative control of environmental pollution caused by toxic metals, and the protection of workers engaged in processing ore in mines, involves the laying down of a set of rules based upon expert recommendations, the prescription of non-offensive levels of air-borne pollutants in the working environment, adoption of international conventions and recommendations concerning the prevention of occupational risks, observation of the codes of practice and guides on prevention, medical supervision, and improvement of the working environment.

A number of acts passed by parliament have been adopted by state governments. For the effective enforcement of these acts, state governments have formulated a number of rules for the guidance of enforcement agencies.

Legislation: The control of toxic metals in air, water, soil, and food, and the protection of the health and safety of workers handling mercury is governed by the following edicts:

1 Smoke Nuisance Acts: these are in force in places such as Calcutta and Howrah, West Bengal (1905), Greater Bombay, Maharashtra (1912) and Ahmedabad, Gujrat (1963). The emissions from iron foundries, and smoke released from various factories dealing with metals are covered by these Acts.

2 Municipalities Act: this is in force in Uttar Pradesh (1916 - section 301, byelaws of the Municipal Board, Kanpur, 1958), Bombay (the Bombay Municipal Corporation Act, 1949 - chapters XII and XVII and sections 310-314) and Gujrat (the Gujrat Municipalities Act, 1963).

3 The Indian Penal Code (1860): chapter XIV, and particularly sections 267 and 268, deal with public nuisances and offences which affect public health and safety.

4 The Code of Criminal Procedures, 1898: sections 133 and 144 provide for the removal of public nuisances.

5 The Mines and Minerals (Regulation and Development) Act, 1957: (Act 67 of 1957). This act came into force on the 1st June, 1958, and superseded the Mines and Minerals (Regulation and Development) Act, 1948.

6 The Workmen's Compensation Act, 1923: this is an important piece of social legislation which provides for payments to workmen, by certain classes of employers, in compensation for accidental personal injury. Of the occupational diseases described in schedule III of the Act, five pertain to metals and/or their compounds. They include poisoning by tetraethyl lead, lead and its alloys, mercury compounds and amalgams, arsenic and its compounds, and manganese and its compounds.

7 The Factories Act, 1948: this is a comprehensive piece of legislation and covers all aspects of the health and safety of workers. In the latest amendment (1976) to the Factories Act, it is mandatory, on the part of factories, to ensure the effective disposal of wastes and effluent and the control of the levels of dust and

fumes which are injurious or offensive to the workers.

8 The Prevention of Food Adulteration Act, 1954: this act regulates the marketing and mixing of non-permitted additives added to food. It was amended in 1976 in order to make it more effective by providing for deterrent punishment, including life imprisonment, of the adulterators. The permissible limits of lead, copper, arsenic, tin, and zinc in foodstuffs have been prescribed in the Act.

EXPOSURE LIMITS

In most industrial operations it is not possible, either to replace the potentially harmful substances, or to avoid their diffusion through the air of the working environment. Therefore, methods have been developed for reducing their concentration to a level which is considered to be inoffensive to the health of the persons exposed. These levels have been established on a somewhat empirical basis; involving either epidemiological data or an extrapolation of the results of toxicological investigations on laboratory animals. In a number of cases, the dose/response relationship has provided a basis for the establishment of exposure limits. Because of certain variations in the determination of the exposure limits, they vary from one country to another. These limits have also undergone continual revision in the light of industrial exposure and scientific research and this has led to a progressive lowering of the limits which were initially considered to be inoffensive.

A number of countries have prescribed exposure limits for chemical substances in the working environment. In India, no systematic attempt has so far been made to arrive at exposure limits for the chemical substances used in industry. Notwithstanding the fact that the determination of exposure limits will have to be based upon socio-economic developments, the state of health of individuals, and climatic conditions, the limits prescribed by other countries are used in India. The Indian Standards Institution (ISI) has issued a number of standards and specifications for the exposure limits to toxic metals. These are based upon the values prescribed by other countries.

Despite the fact that there are a number of laws on the maintenance of the purity of the air, water, and soil, toxic metal pollution continues unabated. The enactment of the Water (Prevention and Control of Pollution) Act (1974) and Rules (1975), and the establishment of central and state Water Pollution Boards have not been able to combat the pollution of bodies of water to any appreciable extent. The Air Pollution Bill, which is shortly to be taken up by parliament for discussion, and the proposed Environmental Protection Law, may help to reduce the pollution load on the environment. More than the formulation of these enactments, their effective implementation, and greater social awareness of the evil of pollution, are two steps which are most essential for combating pollution.

REFERENCES

- Berglund, F.M., Berlin, G., Birke, V., Von Euler, L., Friberg, L., Homstedt, E., Jonsson, C., Ramel, C.S., and Skerfving, A. (1971) - Report of an Expert Group, Nordisk Hygiensk Tidskrift, Supplement

- Brady, G.L. and Cunningham, R.D. (1981) - Ambio, 171-5

- CEK Report (1970) - Environmental Science and Technology, 4, 890

- Clean Air Act Amendments (1970) - PL91-604, USA

- Clean Air Act Amendments (1977) - PL95-95, USA, Sections 323, 405

- Commission of the European Communities (1976) - Official Journal of the European Community, L129, 23-9

- Cotruvo, J.A. (1981) - Science of the Total Environment, 18, 345-56

- FAO/WHO (1974) - WHO Technical Report Series, No.555

- Federal Water Pollution Control Act (1972) - US Government Printing Office, Washington DC

- Haxton, J., Lindsay, D.G., Hislop, J.S., Salmon, L., Dixon, E.J., Evans, W.H., Reid, J.R., Hewitt, C.J., and Jefferies, D.F. (1979) - Environmental Research, 18, 351-68

- Hunt, G.J. and Shepherd, J.G. (1980) - in Radiation Protection: a Systematic Approach to Safety, Proceedings of an International Congress, IRPA, 5th, Jerusalem, 1980 (edited by A.Eisenberg), 141-5, Pergamon Press, Oxford and New York

- Irukayama, K. (1966) - Advances in Water Pollution Research, 3, 153

- Jernelov, H. (1969) - in Chemical Fallout, Charles C.Thomas, Springfield, Illinois, 75

- Krenkel, P. (1979) - Journal of the Water Pollution Control Federation, 51, 2168-88

- National Interim Primary Drinking Water Regulations (1975) - 40 CFR, 141, USA

- Rupp, N.W. and Paffenberger, G.C. (1971) - Journal of the American Dental Association, 82, 401

- Schubert, R.H.W. (1977) - Appraisal of the Ecological Consequences (Ecotoxicology) of Mercury Discharges for the Aquatic Environment: Conclusions of Report ENV/682/77-EN, Brussels, CEC.

- Spence, R.W. (1971) - Journal of Medicine, 285, 971

- Study Group on Mercury Hazards (1970) - Hazards of Mercury: Special Report to the Secretary's Pesticide Advisory Committee, Department of Health, Education, and Welfare, and Environmental Protection Agency

APPENDIX I

ADDITIONAL TABLES

Table I(1) Atomic Abundance of Hg in Chondrites per 10^6 Atoms of Silicon (according to Ehman & Lovering (1967)

Chondrite	Range of Values	Mean Value
Carbonaceous		
Type I	22 - 148	29
Type II	1.6 - 7.2	4.2
Type III	0.62 - 6.6	2.8
Enstatite		
Type A	1.3	1.3
Type B	0.35 - 0.59	0.47
Olivine-hypersthene &		
Olivine-bronzite	0.023 - 0.28	0.10

Table I(2) Mercury Content of Achondrites

Achondrite and Recovery Site	Hg Content (ppm)
ANGRITE	
Angra dos Reis	2.51

continued

Table I(2) (continued) Mercury Content of Achondrites

Achondrite and Recovery Site	Hg Content (ppm)
NAKHLITE	
Nakha	0.23
EUCRITES	
Juvinas	5.01
Moore County	2.74
Nuevo Laredo	0.078
Serra de Mage	0.21
Stannern	0.12
HOWARDITE	
Binda	0.66
DIOGENITE	
Ellemeet	0.43
Johnstown	0.12
AUBRITE	
Norton County	0.054

Table I(3) Mercury in Hypersthene Chondrites (10^{-6}g/g)
[from Reed and Jovanovic (1967)]

Meteorite	Hg_{total}	$Hg_{>450C}$
Peace River	0.015	0.0024
Erymka	0.020	0.00025
Bruderheim	0.044	0.0016
Farmington	0.049	0.0027
Baratta	0.070	0.0014
Canakhale	0.11	0.0027
McKinney	0.19	0.0074

continued

Table I(3) (continued) Mercury in Hypersthene Chondrites (10^{-6}g/g)
[from Reed and Jovanovic (1967)]

Meteorite	Hg_{total}	$Hg_{>450C}$
Holbrok	0.44	0.0037
Harleton	0.44	0.013
Hallingeberg	0.46	0.014
Monze	0.82	0.0075
New Concord	0.89	0.014
Moes	0.92	0.040
Kunashak	1.19	0.0077
Colby, Wise	1.32	0.31
Bjurbole	2.87	0.034

Table I(4) Mercury in Bronzite Chondrites (10^{-6}g/g)
[from Reed and Jovanovic (1967)]

Meteorite	Hg_{total}	$Hg_{>450C}$
Forest City	0.26	0.10
Ehole	0.27	0.028
Pantar-2	0.35	0.12
Pantar-D	0.46	0.017
Pantar-L	0.60	0.0063
Richardton	1.60	0.27
Rose City	4.10	0.23
Breitscheid-D	3.95	0.17
Breitscheid-L	4.47	0.09
Allegan	5.70	0.35
Sindhri	5.84	0.039
Tieschitz	7.22	0.25
Pultusk	7.45	0.58

Table I(5) Mercury in Enstatite and Amphoterite Chondrites (10^{-6}g/g) [from Reed and Jovanovic (1967)]

Meteorite	Hg_{total}	$Hg_{>450C}$
ENSTATITE		
Hvitti	0.80	0.20
Indarch	0.16	0.009
Abee	1.4	0.05
Abee	0.20	0.015
AMPHOTERITE		
Hamlet	0.25	0.11
Soko-Banja	0.84	0.05

Table I(6) Mercury in Carbonaceous Chondrites (10^{-6}g/g) [from Reed and Jovanovic (1967)]

Meteorite	Hg_{total}	$Hg_{>450C}$
Carbonaceous-1		
Orgeuil	213.0	1.8
Orgueil	15.0	0.21
Orgueil	2.4	0.025
Ivuna	4.96	0.39
Carbonaceous-2		
Mighei	4.9	0.23
Carbonaceous-3		
Lance	1.6	0.050
Karoonda	0.80	0.0066

Appendix I

Table I(7) Mercury in Amphoterite Chondrite (10^{-6}g/g)
[from Reed and Jovanovic (1967)]

Meteorite	Hg_{total}	$Hg_{>450C}$
Hamlet	0.25	0.11
Soko-Banju	0.84	0.05

Table I(8) Concentration of Mercury in Meteorites (10^{-6}g/g)
[from Ozerova et al (1973)]

Meteorite	Hg Content
Alexandrovich	120
Bakhmut	90
Bjurbole	62
Bjurbole	200
Dimmitt	1.2
Efremovka	1.5
Elenovka	2.8
Elenovka, magnetic fraction	4.4
Elenovka, non-magnetic fraction	4
Forest City	4.5
Ghubara	2.5
Jelica	74
Kaande (Oesal)	24
Kaande (Oesal)	1700
Krasnyi Ugol	90
Krymka	2
Kunashak, dark	2
Kunashak, light	12
Manych	2.5

continued

Table I(8) (continued) Concentration of Mercury in Meteorites
[from Ozerova et al (1973)]

Meteorite	Hg Content
Nikol'skoe	3.7
Nikol'skoe	2.5
Okhansk	38
Oubari	12
Parnallee	1.5
Pervomaiskh, dark	3
Pervomaiskh, light	2
Pillistfer	20
Plainview	6
Rechki	22
Sartov	19
Saratov, Chondrules	11
Saratov, Troilite	130
Savchenskoe	27
Sevrukovo, dark	5
Sevrukovo, light	4.1
Sevrukovo	50
Stavropol'	28
Tennasilm	140
Tennasilm troilite	1200
Timochin	350
Vavilovka	370
Verkhne-Chirskaia	60
Zhovtnevy	30
Carbonaceous chondrites	
Allende	6
Boriskino	1

continued

Table I(8) (continued) Concentration of Mercury in Meteorites
[from Ozerova et al (1973)]

Meteorite	Hg Content
Grosnaja	7
Mighei	20
Murray	110
Orgeuil	50000
Achondrites Feldspar-free	
Norton County	1.4
Novo Urei	9
Pesyanoe	1.9
Feldspathic (eucrite)	
Chervony kut	1.5
Iron-Stony (Pallasites)	
Brahin	3.5
Brahin Olivine	9.4
Brahin troilite	17
Pallas Iron (Krasnojarsk)	1200
Pallas Iron nickel-iron	40
Olivine	1.2
Stony-Irons (Mesosiderite)	
Budulan	17
Budulan, non-magnetic fraction	2.2
Budulan, magnetic fraction	12
Irons	
Bodaibo	1

continued

Table I(8) (continued) Concentration of Mercury in Meteorites
[from Ozerova et al (1973)]

Meteorite	Hg Content
Boguslavka	3
Canyon Diablo	12
Cape York	4
Chinge	2.8
El'ga, magnetic fraction	2.8
Henbury	2
Niro	3
Santa Catharina	4.3
Santa Catharina troilite	49
Sikhote-Alin	1.8
Sikhote-Alin troilite	23
Sikhote-Alin	3
Tubil	2

APPENDIX II

UNITS

μg/l	(microgram per litre)	Equivalent to parts per billion (ppb) in dilute solution.
mg/l	(milligram per litre)	Equivalent to parts per million (ppm) in dilute solution.
ng/m^3	(nanogram per cubic metre)	Generally used to describe the concentration in the atmosphere. It is approximately equal to 0.001ppb.

APPENDIX III

MEDICAL GLOSSARY

Acrodynia	A painful reddening of the upper and lower limbs.
Amniotic fluid	The fluid in which the foetus floats.
Arterioles	The minute vessels into which the arteries ramify extensively during their course through the body.
Ataxia	A sensory defect causing a loss of the ability to govern muscles, leading to erratic and jerky movements. Muscle power is retained.
Bactericide	Any agent which destroys bacteria.
Bowman's capsule	The part of the kidney which opens into the first portion of the highly convoluted proximal tubule and collects filtrated fluid from the capillaries (figure IIIB).
Cerebellum	The part of the brain which lies behind and below the cerebrum. Its chief functions are the coordination of fine voluntary movements and the control of posture.
Chelates	A class of metal-organic compounds in which the metal atom or ion is held between two atoms of a single molecule (the ligand). The chelating ligand usually attaches itself to a metal ion via nitrogen, oxygen, or sulphur atoms.
Chromosome	A thread-like body in the cell nucleus which divides during mitosis and carries hereditary factors (genes).
Duodenum	The fixed curved first part of the small intestine.
Eczematous	A type of dermatitis which begins with erythema followed by the appearance of vesicles.
Enterohepatic disease	An ailment affecting the intestines and liver.
Epidemiology	The scientific study of the spread and distribution of diseases.

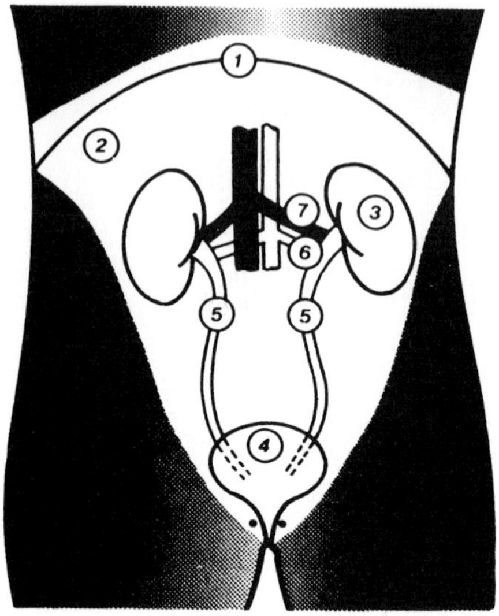

Figure IIIA
Human Excretory System
(1 - diaphragm, 2 - abdominal cavity, 3 - left kidney, 4 - bladder, 5 - ureter, 6 - renal artery, 7 - renal vein)

Epithelial tissue	The covering of the body's ducts, vessels, and cavities.
Erythematous	Pertaining to erythema.
Erythema	A reddening of the skin due to an allergic skin reaction.
Erythrocytes	The normal red-cells of circulating blood.
Fungicide	Any agent which destroys fungi.
Glomerulus	A coil of minute arterial capillaries, held together sparingly by connective tissue. It invaginates the entrance to the uriniferous tubules in the renal cortex (figure IIIB).
Ileum	The lower three-fifths of the small intestine, lying between the jejunum and caecum.
In vitro	Experiments carried out on tissue samples separated from living animals (literally - in glass).

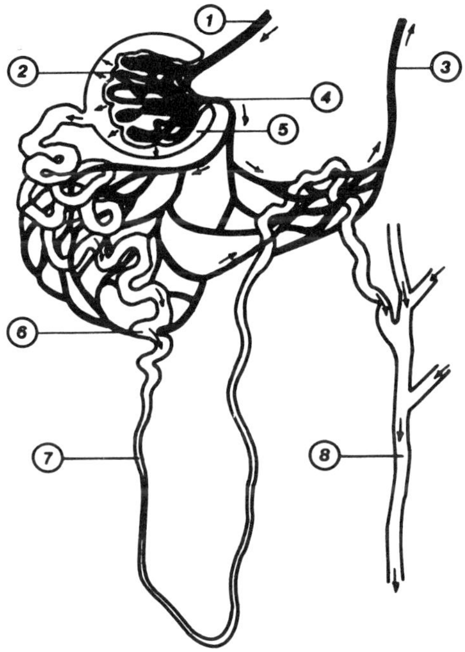

Figure IIIB
Simplified Structure of a Nephron
(1 - afferent vessel, 2 - glomerulus, 3 - vein,
4 - efferent vessel, 5 - Bowman's capsule, 6 - renal tubule,
7 - loop of Henle, 8 - collecting tubule)

In vivo	Experiments carried out on living animals (literally - in life).
Jejunum	The part (about 2.5m in length) of the small intestine which lies between the duodenum and the ileum.
Lipid	The fatty inner layer of the cell membrane.
Loop of Henle	A U-shaped renal tubule in the nephron (figure IIIB).
Lymphocyte	A protein-rich type of cell which is formed in the lymph nodes.
Lysol	A saponified solution of cresol, used as a disinfectant.
Mutagenic	Produced by gene mutation or chromosome aberration.
Neonate	Pertaining to a baby up to one month old.

Neurasthenic	A rare nervous condition characterised by lassitude, inertia, fatigue, and loss of initiative.
Neurotoxic	Poisonous and destructive to nerve tissue.
Parathesia	Any abnormality of sensation.
Pink disease	See Acrodynia.
Renal tubules	These handle the selective re-absorption of substances from the glomerular filtrate until the latter reaches the composition of urine (figure IIIB).
Stomatitis	Inflammation of the mouth.
Symphysis	A fibrocartilaginous union of bones.
Teratogenic	A disruption of foetal growth, leading to malformation, due to the action of drugs taken during pregnancy.
Trophic	Pertaining to nutrition.
Ureter	A tube running from each kidney to the bladder (figure IIIA).
Urethra	The tube through which urine is discharged from the bladder.

APPENDIX IV

GEOLOGICAL GLOSSARY

Abyssal	Deep-seated. For example, abyssal rocks, abyssal faults.
Achondrite	A form of stony or carbonaceous meteorite which is free of spherical chondrules. It resembles a terrestrial rock.
Acid rock	An igneous rock whose bulk composition consists of more than 66% of SiO_2. Granite is one of the most common acid rocks.
Aerosol	A dispersion of solid or liquid particles in air. For example, smoke.
Albitisation	The development of albite plagioclase (Ca,K,Na aluminosilicate) in a rock at the expense of pre-existing mineral by the introduction of Na ions via a process known as metasomation.
Algae	Sea-weeds, pond-scum, etc. These are thallophytes which are characterised by the fact that they contain chlorophyll and are capable of obtaining their food via photosynthesis.
Alkali rocks	Igneous rocks which contain relatively large amounts of alkalies (i.e., Na and/or K). For example, alkali basalt.
Aluminosilicates	A large number of clay minerals are aluminosilicates. For instance, kaolinite, montmorillonite, etc. They have the general composition: [Al_2O_3, $nSiO_2$, mH_2O].
Amphibolite	An important basic rock which is produced by metamorphism and is composed essentially of a dark-coloured mineral called amphibole. This is a metasilicate with the composition, $RSiO_3$, where R represents Al, Ca, Fe, K, Mg, Na, or Ti.

Andesite	A basic igneous rock which is composed mainly of plagioclase pyroxene or hornblende, biotite, and some minor ore minerals.
Antimonite	A mineral containing antimony sulphide, Sb_2S_3.
Aragonite	A calcium carbonate ($CaCO_3$) mineral like calcite. It has an orthorhombic structure and is commonly needle-like in form. It is formed at 80 to 100°C and is relatively unstable at normal pressures and temperatures.
Argillites	Dense, fine-grained detrital sedimentary rocks which are commonly composed of clays, mud, silt, quartz, and carbonates.
Arsenopyrite	Iron arsenosulphide, FeAsS.
Barytes	A mineral having the composition, $BaSO_4$. It has an orthorhombic structure and is commonly found in sedimentary rocks.
Basic	See ultrabasic.
Benioff zone	A sloping surface of seismic activity which is characteristic of island-arc systems. The Benioff zone meets the Earth's surface close to an ocean trench and dips below the island arc.
Biotite	A brown/black mica containing iron and magnesium.
Cambrian	See table IV(1).
Carboniferous	See table IV(1).
Chalcedony	A cryptocrystalline variety of silica.
Chalcopyrite	Copper iron sulphide mineral, $CuFeS_2$.
Chlorite	A group of minerals with the phyllosilicate structure, related to talc.
Chondrites	Stony meteorites containing small spherical aggregates of minerals known as chondrules.
Clarke	The ratio of the concentration of a given element in a rock to the average concentration of that element in the Earth's crust.

Clay minerals	A group of silicate minerals with a layer structure, formed by the weathering of silicate rocks. It includes the minerals: kaolinite, illite, montmorillonite, and vermiculite.
Connate water	Water trapped in sediments at the time of deposition.
Cumulus phase	Aggregates of crystals formed by settling in a molten magma.
Devonian	See table IV(1).
Diabase	An igneous rock which is composed mainly of Ca-rich plagioclase and clinopyroxene.
Diapir	A structure in which the core of an anticline breaks through overlying rocks.
Diastrophism	The deformation of large masses of the Earth's crust to form mountains, etc.
Dolerite	A dark-coloured igneous rock which is basic in nature and resembles basalt, but is composed of rather larger crystals.
Dunite	A rock which is composed essentially of olivine, $(Mg,Fe)_2SiO_4$.
Dyke	A more or less vertical wall-like mass of igneous rock which cuts across the rocks into which it is intruded.
Eclogite	A metamorphic rock formed at high temperatures and pressures, and comprising garnet, pyroxene (omphacite), amphibole, etc.
Effusives	Rocks which have been formed by the effusion of molten magma, from within the earth, through an opening.
Endogenetic	A term which is used to describe processes which originate from within the Earth.
Enstatite	A pyroxene having orthorhombic symmetry (i.e. orthopyroxene) and the composition, $MgSiO_3$.
Epigenetic	A term which is used to describe processes which occur within the Earth.
Epithermal	Pertaining to low-temperature (100 - 200°C) hydrothermal processes.

Fault	A dislocation of a rock formation, in which one side has moved relative to the other.
Feldspar	Major rock-forming silicate minerals comprising four main groups: potassium feldspar or orthoclase ($KAlSi_3O_8$), sodium feldspar or albite ($NaAlSi_3O_8$), calcium feldspar or anorthite ($CaAl_2Si_2O_8$), and barium feldspar ($BaAl_2Si_2O_8$).
Forsterite	A Mg-rich olivine, Mg_2SiO_4.
Fumerole	A hot spring which gives off gases.
Gabbro	A basic igneous rock composed mainly of plagioclase and clinopyroxene.
Glaucophane	A mineral found in metamorphic rocks such as schists, and in igneous rocks such as Na-rich granites.
Gneiss	A metamorphic, crystalline, and coarse-grained rock which resembles granite but exhibits a more or less banded arrangement of the constituents.
Granite	A coarse-grained igneous rock which consists essentially of quartz (20 - 40%), Na-K feldspar, and very often a mica: biotite and/or muscovite. It is formed by the slow cooling of an acid magma.
Granodiorite	A coarse-grained acid plutonic rock containing quartz, plagioclase, orthoclase, and minerals such as biotite and hornblende.
Harzburgite	An ultrabasic rock consisting of olivine and orthopyroxene.
Hydrothermal processes	Any process associated with igneous activity involving heated or superheated water.
Hypersthene	An orthopyroxene mineral with the approximate composition, $(Mg,Fe)SiO_3$.
Hypothermal	Pertaining to high-temperature (300 - 500°C) hydrothermal processes.
Igneous rock	A rock formed by the solidification of molten magma.
Island arc	A curved chain of islands with the convex (outer) side of the curve facing the open ocean. These are regions where deep-focus earthquakes occur. The islands exhibit volcanic activity.

Kaolinite	A group of clay minerals with the formula, $Al_4Si_4O_{10}(OH)_8$.
Kimberlite	A thoroughly brecciated peridotite (an ultramafic rock) which consists essentially of mica, bronzite, and chrome-diopside.
Kupferschiefer deposit	The main copper deposit in Mansfeld, Germany. This is a renowned sedimentary base-metal sulphide deposit.
Leucocratic rocks	Light-coloured igneous rocks which consist of minerals such as quartz, feldspar, etc.
Lineament	A large-scale linear topographic feature which reflects an underlying structural feature.
Mafic	See ultramafic.
Melanocratic rocks	Dark-coloured igneous rocks which consist of ferromagnesian minerals.
Mesozoic	See table IV(1).
Meteorite	Extraterrestrial material which falls to earth. It can be described as, chondritic, achondritic, stony-iron, or iron, depending upon the presence or absence of chondrules, and the amount of silicate minerals and iron alloys present.
Mid-oceanic range	A range in the middle of an ocean, e.g., the mid-Atlantic ridge.
Mississippi Valley type	Epigenetic stratified sedimentary base metal sulphide deposits.
Montmorillonite	See kaolinite.
Neritic	Related to shallow seas, especially those over the continental shelf.
Olivine	A mineral having the composition, $(Mg,Fe)_2SiO_4$, and a member of the series of solid solutions ranging from Mg_2SiO_4 (forsterite) to Fe_2SiO_4 (fayalite).
Ophiolite	A group of basic and ultrabasic rocks which occur as lavas and minor intrusions.
Ordovician	See table IV(1).

Table IV(1)
Geological Time Scale

			TIME SCALE (M years)
CENOZOIC	QUATERNARY	HOLOCENE	0.01
		PLEISTOCENE (glacial)	2
	TERTIARY	PLIOCENE	10
		MIOCENE	25
		OLIGOCENE	40
		EOCENE	60
		PALAEOCENE	65
MESOZOIC		CRETACEOUS	135
		JURASSIC	180
		TRIASSIC	230
PALAEOZOIC		PERMIAN	280
		CARBONIFEROUS	345
		DEVONIAN	400
		SILURIAN	440
		ORDOVICIAN	500
		CAMBRIAN	600
PRE-CAMBRIAN			2000

Orogeny — The process of mountain building. Deformation folding and thrusting are characteristic features of orogenics and are accompanied by the intrusion of igneous rocks.

Palaeozoic — See table IV(1).

Permian — See table IV(1).

Phanerozoic	The period between Cambrian and Quaternary, during which sediments were deposited which contained abundant animal and plant remains.
Phonolite	An undersaturated (with respect to the silica content) material containing minerals such as nepheline, etc.
Pillow lava	Lava whose surface resembles pillows.
Plagioclase feldspar	A series of triclinic feldspars with albite ($NaAlSi_3O_8$) and anorthite ($CaAl_2Si_2O_8$) as the end-members.
Plate tectonics	A theory which assumes that the Earth's surface is covered by a number of fragmented rigid pieces of crust (plates) which move over the upper mantle. There are seven large plates, known as the African, Eurasian, Indo-Australian, Pacific, North American, South American, and Antarctic.
Pleistocene	See table IV(1).
Pluton	A cylindrical igneous mass produced by low-temperature intrusion.
Pneumatolysis	Changes produced by the hot gases given off by magma in the later stages of cooling and solidification.
Podsol	Soil formed in a cool temperate climate with a strongly developed leached zone.
Pre-cambrian	See table IV(1).
Propylitisation	The hydrothermal alteration of a fine-grained basic igneous rock into a mass of secondary minerals such as epidote, chlorite, quartz, carbonates, and sericite mica.
Pyrite	Iron sulphide mineral, FeS_2.
Pyroxenes	An isomorphous group of silicate minerals with Si_2O_6 single chains. Enstatite, $Mg_2Si_2O_6$, bronzite, $(Mg,Fe)_2Si_2O_6$, and hypersthene, $(Mg,Fe)_2Si_2O_6$, are members.
Quartzite	A rock which is composed mainly of quartz.
Quaternary	See table IV(1).

Regolith	A layer of fragmented and non-lithified detritus which can overlay a planetary surface, as in the case of the moon.
Rheomorphism	Plastic or viscoplastic deformation of rocks involving partial melting.
Rhyolite	A fine-grained glassy acid volcanic rock with a similar composition to that of granite.
Rift zone	A zone which is bounded by faults belonging to the major tectonic features of the crust. An outstanding example is the East African rift zone which stretches 4000km from Syria to the Zambesi.
Schist	A metamorphic rock which is composed of minerals and is compressed to give a flaky appearance. Mica is often present.
Sea mount	A mountain under the sea which rises from the floor of the ocean but does not reach sea level. All are of volcanic origin.
Sedimentary rock	A rock which is formed by the consolidation of sediments transported by water or air. For example, sandstone and limestone.
Sericitisation	The formation of the low-temperature mica-type mineral, sericite.
Serpentine	A silicate formed by hydration of olivine and pyroxene.
Serpentinite	An ultrabasic rock which is composed of the mineral, serpentine.
Silurian	See table IV(1).
Slate	An argillaceous or clay rock which fractures along planes (cleavage) under the pressures produced by metamorphism.
Solfataric action	The action of fumeroles giving off sulphurous vapours.
Sphalerite	Zinc sulphide mineral, ZnS.

Subduction zone	The region in which a lithospheric plate is subducted (forced down) into the upper mantle. As it moves into the mantle, the plate is heated and, at a depth of between 100 and 300km, it is partially melted (see Benioff zone). The movement of the lithospheric plate is thought to be the cause of the earthquakes which occur in island arc regions.
Syngenetic	Meaning formed at the same time; synonymous with co-genetic.
Tectonics	The study of the major structural features of the Earth's crust or the broad structural features of a region.
Tholeiite	A basalt consisting of calcic plagioclase feldspar and pigeonite (a pyroxene) with interstitial glass.
Troilite	A mineral which is composed of FeS and is a common associate mineral in iron, and most stony, meteorites.
Tuff	A pyroclastic rock formed by the settling and consolidation of the volcanic ash ejected during volcanism, together with other rocks.
Ultrabasic	Rocks which are composed essentially of minerals such as olivine, pyroxene, etc.
Ultramafic	See ultrabasic.
Upper mantle	The upper part of the Earth's mantle, which lies at a depth of between 35 and 2900km. That is, between the Mohorovicic and Gutenberg discontinuities.
Uralite	A product of the low-grade metamorphism of basic igneous rocks, producing amphiboles in place of pyroxenes.
Wehrlite	An ultrabasic rock consisting of olivine and augite.
Xenolith	An inclusion of pre-existing host rock in an intrusive igneous rock.
Zeolites	A group of hydrous aluminosilicate minerals containing sodium, potassium, calcium, and barium.
Zoisite	A type of epidote having the composition, $Ca_2Al_3O.Si_2O_7.SiO_4(OH)$.

LIST OF FIGURES

LIST OF TABLES

List of Tables

AUTHOR INDEX

SUBJECT INDEX